The Republic of Labor
*Philadelphia Artisans
and the Politics of Class,*
1720–1830

Ronald Schultz

New York   Oxford
Oxford University Press
1993

Oxford University Press

Oxford   New York   Toronto
Delhi   Bombay   Calcutta   Madras   Karachi
Kuala Lumpur   Singapore   Hong Kong   Tokyo
Nairobi   Dar es Salaam   Cape Town
Melbourne   Auckland   Madrid

and associated companies in
Berlin   Ibadan

Published by Oxford University Press, Inc.,
200 Madison Avenue, New York, New York 10016

Library of Congress Cataloging-in-Publication Data
Schultz, Ronald, 1946–
The republic of labor : Philadelphia artisans and the politics
of class, 1720–1830 / Ronald Schultz.
p. cm.   Includes bibliographical references and index.
ISBN 0–19–507585–4
1. Working class—Pennsylvania—Philadelphia—History.
2. Artisans—Pennsylvania—Philadelphia—History.
I. Title.
HD8085.P53S38   1993   305.9′6—dc20
92–21082

9 8 7 6 5 4 3 2 1
Printed in the United States of America
on acid-free paper

For Sara and Tristan

Acknowledgments

No author works alone and I have accrued many debts in the course of writing this book. My most enduring debt is to three of my graduate teachers: Robert Brenner, who taught me about early modern England and how to ask the right questions; James Henretta, who eased the transition from English to American history; and Gary Nash, who revealed the multiform world of early America with commitment and grace. Without them, graduate school would have been neither as exciting nor rewarding as it was. I thank them for this and much else besides.

This book began as a dissertation and would have retained all too many of its stylistic quirks and underdeveloped themes without the careful readings of Gary Nash, Bruce Laurie, Milton Cantor, and George Rúde. I thank them all, but especially Gary Nash, who urged me to make religion a larger part of the story. Sara Kenyon was kind enough to read several portions of the manuscript at a time when her own teaching and writing were pressing enough. As a non-historian, she helped me to make my thoughts clearer. My thanks as well to the anonymous referees who offered valuable suggestions and helpful encouragement in just the right proportions.

Without doubt, my greatest debt is to Gary Nash. For more than a decade, Gary has been a source of inspiration, a model to emulate, a tough and helpful critic, and, in the end, a friend. I'm not sure I can ever thank him enough, but I know this book is the kind of thanks he values most.

This book could not have been written without the assistance of numerous archivists and librarians and I've been fortunate to work with some of the very best. At the Historical Society of Pennsylvania Linda Stanley helped far beyond the call of duty, as did Phil Lapsansky at the Library Company of Philadelphia. The other staff members of both institutions made my many research trips to Philadelphia more than worthwhile. At the University of Wyoming, the librarians and staff of the William Robertson Coe Library have helped by making that institution a cordial and productive place to work. My special thanks to the many people of the Inter-Library Loan Department, who have dealt with my voluminous and often obscure requests with dogged efficiency and unfailing good humor.

For financial assistance, I would like to thank the National Endowment for the Humanities, the Division of Basic Research of the University of Wyoming, and for help in purchasing photographs for this book, the Department of History, also at the University of Wyoming.

To my colleague, Phil Roberts, very special thanks for compiling the index.

At Oxford University Press, I am indebted to Karen Wolny and Leona Capeless for guiding this book through the process of publication and to Stephanie Sakson for her keen eye and expert suggestions as copy editor.

Brief passages in Chapters 1 through 5 first appeared in "The Small Producer Tradition and the Moral Origins of Artisan Radicalism in Philadelphia, 1720–1810," *Past and Present: A Journal of Historical Studies* 127 (May 1990), pp. 84–116. (World Copyright: The Past and Present Society, 175 Banbury Road, Oxford, England). My thanks to the Society for permission to use that material here.

Last, and most important, my wife, Sara Kenyon, and our son, Tristan Kenyon-Schultz, have been constant companions throughout the very long process of researching and writing this book. Through the years, they have taught me much about what is important in human relationships and have also helped me to keep the lives of Philadelphia artisans in something approaching a reasonable perspective. I thank them for their forbearance and, especially, for their love.

Fort Collins, Colorado   R. S.
September 1992

Contents

Prologue *xi*

1. The Anglo-American Radical Traditions
   of the Seventeenth and Eighteenth Centuries   3

2. Property Rights and Community Rights:
   The Politics of Popular Revolution, 1765–1779   37

3. Reaction and Restoration:
   Artisan Politics in Crisis, 1780–1789   69

4. Hegemony and Counter-Hegemony:
   The Rebirth of Popular Radicalism, 1790–1795   103

5. An Apprenticeship to Class, 1796–1810   141

6. Confronting Industrialism:
   Artisan Politics and Philadelphia's
   Socialist Tradition, 1810–1820   181

7. Making the Republic of Labor:
   The Workingmen's Movement and American
   Working-Class Consciousness, 1820–1830   211

Epilogue   After the Workingmen's Party:
Labor and the Politics of Class   235

Notes   239

Index   285

Prologue

. . . there are many things which we know better and feel much more
strongly than the richer, softer-handed classes can know or feel them. . . .

GEORGE ELIOT, *Address to Workingmen* (1868)

"To the working class . . . the system of profit . . . is, and will
continue to be so, as long as it continues at all, an iron chain of
bondage."[1] With these words, William Heighton, an obscure journeyman cordwainer, announced the birth of the American working
class. Addressing the members of Philadelphia's journeyman trade
societies, Heighton spoke of subjects already well known to his audience. Lecturing in 1827, his call for working-class unity surprised
few of his listeners, for they knew that Pennsylvania's port city had
been a center of craft organization since at least the 1780s. Likewise,
his plan for the formation of a workingman's political party was
readily accepted, for most journeymen knew that Philadelphia artisans had been actively politicized for more than a century. Even his
call for artisan socialism raised few eyebrows, for many knew that
Philadelphia had supported an indigenous socialist movement since
the turn of the century. Heighton's speech, then, was both a call to
action and a summing up of traditions with deep historical roots.

Historians have been slow to appreciate the things that William
Heighton and his fellow craftsmen knew. Without exception they
have written about the rise of the working class as a nineteenth-
century phenomenon, most recently as a social formation precipi-
tated in the confrontation between industrial transformation and
republican ideology.[2] Yet as Heighton and his audience understood,
the roots of the American working class ran deep into the eighteenth
century and beyond. If historians have come to view the creation of
the Mechanics' Union of Trade Associations and the Workingmen's

Party as the dawn of the American labor movement, they have forgotten that the events of 1827 and 1828 were an end as much as a beginning. For the American working class made itself, not in 1827, but in the course of a century of economic and political struggle that saw independent colonial artisans transformed into skilled but dependent workingmen.

This book is a study of the forces that shaped the experiences of Philadelphia workingmen in this forgotten era. The first of these forces was a tradition of artisan thought that connected early American craftsmen with the moral traditions of their English predecessors and contemporary counterparts. The second was a tradition of popular politicization that was unique in the Atlantic world. It was through the twin perspectives of these popular traditions that Philadelphia workingmen interpreted what was, perhaps, the largest force of all: the economic and social changes brought about by early industrial capitalism. Between 1720 and 1830 these forces came together in the streets, workshops, and meeting rooms of Philadelphia and resulted in the creation of the nation's first working class.

It was a unique creation, this working class, for neither industrial capitalism nor modern class divisions came easily to Philadelphia. Long the center of American craft production, the transition to capitalism came to Philadelphia, as it came to the rest of nineteenth-century America, as the demise of a distinctive class formation and a unique way of life. Welcomed by some and condemned by many others, early American capitalism took root among the shattered fragments of a system of independent proprietorship and family production that had lasted, in some areas, nearly two hundred years. In its place rose the system of class relationships with which we are familiar to this day: an alignment that pitted manufacturing and financial capitalists against an increasingly dependent class of direct producers.

It is only by viewing the rise of capitalism as the simultaneous destruction of a deeply valued system of production that we can understand the true nature of working-class formation in the Delaware city. Artisans created a workers' movement of unusual poignancy in Philadelphia not only because they felt themselves exploited, but because they had something very real to defend. And when their early attempts failed, they continued to combat multiplying relations of dependency by creating a culture that combined their old class ideals with the collective experience of political mobilization. Throughout a century of transition and class formation, Philadelphia artisans and workingmen never totally abandoned hope for the restoration of the economic and social independence of the craftsman. But as they increasingly came to see, the key to their success lay in politics and the workingmen's ability to link producer traditions and political action. This is why the answer of Philadelphia workingmen to the new social system of industrial capitalism was, in the end, the creation of a republic of labor.

The following pages tell the story of the artisans' struggle chronologically: the way in which they experienced the transformation of their lives. For this reason, I have departed from the common practice of discussing the decline of craft production and the rise of capitalist manufacturing in separate chapters and have, instead, presented these issues as they confronted Philadelphia artisans: in the medium of time. In this way, it is hoped that the reader will be drawn closer to the experience of the tailor, carpenter, and shoemaker for whom capitalism was not an economic system, but a loss of craft prerogative, a threat to competency, and the condescension of wealthy men who lived from the fruits of their labor.

Chapter 1 begins by tracing a uniquely artisan tradition of small producer thought from its origins in fifteenth-century England, through its development during the English civil war, and then to its transmission, in the minds of emigrating Quaker artisans, to the small producer haven of Pennsylvania. It was to this artisan moral tradition, encompassing an ethic of community, equality, competency, and the value of labor, that Philadelphia political leaders directed their eighteenth-century appeals for popular support and in the process created the city's tradition of popular coalition politics.

Chapter 2 follows this tradition of political coalition into the American Revolution where the bonds between middle-class leaders and city mechanics were tested and found wanting. The turning point was the price-control movement of 1779 which revealed a fundamental difference between eighteenth-century popular leaders, who elevated absolute property rights above the needs of the community, and rank-and-file craftsmen who retained the values and community-mindedness of the small producer tradition. It was a division that would plague popular politics throughout the years of economic transition and class formation portrayed in this study.

Chapter 3 surveys the effects of the wartime breakdown of coalition politics on the emerging craft consciousness of the post-Revolutionary era. Following the Revolution, workingmen divided: some maintained their ties with middle-class radicals while most transferred their allegiances to the mainstream Republican and Constitutionalist parties as their immediate interests dictated. The most important result of this period of political fragmentation, however, was a turning inward on the part of many Philadelphia craftsmen and the development of a notion of laboring-class self-sufficiency. During the 1780s, the traditional craft system came under increasing attack as growing numbers of city masters began the process that would make them capitalist employers. As a result, Philadelphia journeymen formed their own craft organizations and, for the first time in the Anglo-American world, issued calls for an independent laboring-class politics.

Chapter 4 charts the development of an anti-Federalist movement in Philadelphia. From the beginning, middle-class Democratic-Republicans realized that their success rested on the support of the city's mechanics, and they canvassed this support by appealing, once again, to the artisan's small producer values. At the same time, however, the growth of trade societies among Philadelphia journeymen signaled an ever-widening division between masters and journeymen that threatened to destroy the very basis of the Jeffersonian coalition.

With Chapter 5 the story moves into the nineteenth century and with it a decade of troubled, but ultimately successful alliance between Philadelphia Democratic-Republicans and the city's workingmen. Much of the credit for this success belonged to the ideas and leadership of William Duane, editor of the nation's pre-eminent Jeffersonian newspaper, the *Aurora*, and to the organizing abilities of the Northern Liberties politician, Dr. Michael Leib. Together the two men forged a laboring-class wing of the Democratic-Republican party that was the cornerstone of party strength in Philadelphia. Yet despite the palpable success of their efforts among city workingmen, after 1800 Philadelphia entered a period of early industrialization that was to weaken, and eventually doom, this multi-class coalition.

Chapter 6 begins with the demise of the Jeffersonian coalition in the election of 1810. This election, which divided the Democratic-Republican party into two factions, one representing merchants and master manufacturers and the other ordinary workingmen, helped to create a new awareness of class among Philadelphia journeymen. This emerging sense of class was reinforced by the rise of an indigenous socialist tradition in Philadelphia which introduced the idea of economic exploitation to city workingmen and, in so doing, took the small producer tradition to its radical limits. Coupled with the devastating effects of the Panic of 1819, the stage was set for the rise of a conscious working-class movement.

Chapter 7 brings the story to a close by following the peripatetic campaign of William Heighton, a young journeyman shoemaker and advocate of the English socialist John Gray. During the mid-1820s, Heighton canvassed the city attempting to draw together the disparate threads of rational Christianity, artisan socialism, and trade society organization into a plan that would create a republic of labor. After two years of almost ceaseless activity, he finally met with success: in 1827 Philadelphia journeymen formed a city-wide central trade union and, early the following year, organized a workingmen's political party. The result was the birth of America's first working class.

The Republic of Labor

CHAPTER 1

The Anglo-American Radical Traditions of the Seventeenth and Eighteenth Centuries

"A PEOPLE is travelling *fast* to destruction, when individuals consider their interests as distinct from those of the public."[1] It was with these words, borrowed from John Dickinson's *Letters*, that Charles Thomson attempted to pull Philadelphia's merchants into the non-importation movement of 1768. Community protest against the Townshend duties had already swept Boston and New York; only Philadelphia remained among the major colonial ports open to British imports. Thomson, a sometime schoolteacher and shopkeeper, was an early leader of the radical-popular movement in Philadelphia, and it was in this capacity that he sought to bring the full force of community sentiment against the indolence and dissimulation of the city's merchants. "Now is the time for exertion," he insisted. "If not, when shall we exert ourselves? When all the *intolerable* evils of arbitrary power shall have *preyed* upon our vitals?"[2] He ended his argument with a final appeal for all Philadelphians to "Unite in the *good* cause."[3] Significantly, Thompson signed himself "A FREEBORN AMERICAN."

In its rhetorical appeal to Whig anti-absolutism as well as in its gesture toward the power of community norms, freeborn rights, and the "good old cause" of the English civil war, Thomson's address reminds us of two facts that are crucial in any account of the rise of the working-class movement in Philadelphia. First, it reminds us of the importance of English moral traditions during the first century of class formation. And not only the republican traditions of the educated classes but also the indigenous traditions of artisans and workingmen. Second, it reminds us that the working-class movement was always a part of the politics of the age. In Philadelphia, politics was seldom the unchallenged property of an elite, and opposition parties and multi-class coalitions early became the hallmark of the city's political life.

Artisan traditions and popular politics were the ground from which Philadelphia's working class emerged in the early nineteenth century. To understand how this happened, our story must begin not in the Quaker City, but in the plebeian culture of early modern England. For it was there, among rural clothworkers and London craftsmen, that artisan culture first took its modern form. And it was there, during the 1640s, that this culture first merged with the political radicalism of the Levellers to create a prototype for artisan political activism in the Quaker City. And it was also there, among the Quakers and radical sectarians of the seventeenth century, that an imprint of religious radicalism was fixed onto the spirit of artisan culture.

I

Popular moral traditions seldom reveal their origins in any precise or unequivocal way. It is only when they attract official attention or voice objections that obtrude into the everyday functions of elite rule that they make a mark on the historical record. Such is the case with the tradition of small producer thought that emerged from the rural workshops of sixteenth-century English clothworkers and London artisans to become the distinctive credo of artisans throughout the Anglo-American world. Like other aspects of artisan culture, this small producer tradition was bound up in the oral culture of the shop, tavern, and artisan neighborhood and offered itself to public scrutiny only during sporadic periods of political and economic stress. But when English and American artisans did enter into larger public debates, they did so wielding their own distinctive idiom of small producer thought. It is to that idiom that we first turn.

Perhaps the oldest notion in the artisan's lexicon was the sense of pride that craftsmen derived from the social utility of their labor. We have lost much of this feeling today, but for an artisan, work was never merely physical labor performed for a requisite number of hours each day. Whether laying the keel of an oceangoing merchantman or fashioning something as simple and prosaic as a shoe, craftsmen looked upon useful labor as a moral and social, as well as an economic, act. This stress on the social and ethical qualities of labor owed a great deal, of course, to the nature of the pre-industrial community, which depended on the artisan's skilled labor for the provision of virtually all of its goods and services. The artisan was literally at the center of community life, making up, as one observer put it, the "centre of population, the axis of society."[4] This was not self-serving rhetoric, but simple truth. Before the rise of widespread markets and advanced manufacturing processes, the everyday functioning of all communities rested on the artisan's vital contribution to the local economy. And befitting his pivotal position, the artisan felt himself entitled not only to a comfortable and secure existence but to the respect and esteem of the community which he served.

If skilled labor formed the productive underpinning of community life, it followed that the artisan was the equal of any man. In an age when the ownership of property alone conferred political rights, artisans grasped this idea and turned it abruptly around. In place of the tangible property of wealthy merchants and large landowners, artisans substituted the skills they carried in their heads and hands, declaring that skill was, like land and stock, a form of property that guaranteed independent political judgment. As one Connecticut mechanic proclaimed, skill was not only the accumulated "genius of the people," but "that property which is . . . the product and natural profits of the quantity of industry and economy bestowed [by] seven years strict application to a particular science."[5] Like the farmer who acquired property rights through the labor he and his family invested in the land, the artisan acquired equivalent rights through the labor he invested in a long period of apprenticeship. "Such property and such abilities," argued Norwich's Walter Brewster, "are acquired and not natural, they are actually the purchase of industry in the strictest sense."[6] For Brewster and other

Anglo-American craftsmen, the property rights of skill made the artisan the political equal of any merchant or commercial landholder.[7]

This argument, with its Lockean and legalistic overtones, derived from the interchange of artisan and educated cultures. Yet for serviceable ideas of social equality, artisans had only to look to the everyday operations of their shops. From the day he signed on as an apprentice, the young artisan witnessed everyday democracy in action. Despite the master's formal control over the workplace, shop tasks, work load, production speed, and more were decided by the formal rituals of the trade as well as by ad hoc convocations of masters and shopmates.[8] By the time they completed their terms, junior craftsmen not only understood the importance of their skills, but they expected to render independent judgments in shop matters, in their trade societies, and in the wider world as well.

Underwriting the craftsman's pride in his social usefulness and supporting his claim to social equality was the notion of competency. The achievement of a competency was the goal of every artisan and he invested his youth in apprenticeship to attain it.[9] For countless generations of craftsmen, a competency was the prospect of moderate prosperity that a life of skilled and useful labor rightfully commanded. This aspect of competency was well captured in a popular verse directed at late eighteenth-century artisans:

> Let me, O God, my labours so employ,
> That I a competency may enjoy;
> I ask no more that my life's wants supply,
> And leave their due to others when I die;
> If this thou grant (which nothing doubt I can),
> There never liv'd or di'd a richer man.[10]

But if competency denoted an artisan's financial independence, it always meant more to him than simple economic expectation. Implicit in its notion of social responsibility was an unwritten covenant between the artisan and his community. In offering his fellow citizens a lifetime of productive labor, the skilled artisan expected to receive the respect of his community and a life free from protracted want in return. The meaning of competency thus went well beyond mere pecuniation: at its core, competency was an expectation of middling status in the community, an acknowledgment of economic independence that brought with it small comforts, a few luxuries, and an abiding sense of self-esteem.

An intense commitment to community completed the small producer tradition. Like their feelings of social equality, artisans learned the meaning of community from the society of their shops. With few exceptions, pre-factory artisans labored in small shops where a master worked with one or two journeymen and a like number of apprentices. Even the few who worked in larger-scale settings such as shipyards, ropewalks, or metal works labored in tight-knit work gangs of ten to twelve men. Working life in these settings was necessarily intimate and goverend by overarching norms of mutuality and cooperation. Structured like a family, the shop or work gang formed a little commonwealth where individual efforts were ultimately regulated by the collective well-being of the whole. Transferred into the wider world, artisans imagined their community as an association of individuals laboring separately for the ultimate benefit of all. For them, a well-run society was much like a well-regulated trade: it required the subordination of individual self-interest and personal acquisitiveness to the collective well-being of the entire citizenry.

II

The ideas that made up the small producer tradition were already old by the eighteenth century, and for their origins we must look to the declining years of English feudalism. The manorial economy never recovered from the Black Death that ravaged England in the fourteenth century, and in its place arose a rural economy based on capitalist farming and the large-scale pasturage of sheep.[11] By the early sixteenth century, the cheap food and plentiful wool supplied by this economy had made possible the creation of a widespread provincial outwork system that produced woollen cloth for England's thriving export trade.

At the center of this trading system were hundreds of rural craftsmen who worked up shearings into semi-finished cloth and shipped their week's work to the London merchants who controlled the English cloth trade. These rural craftsmen were, along with their fellow artisans of the City of London, the most economically and socially advanced producers of western Europe. They were also the most ideologically advanced class in English society.[12] Freed from the ties of feudal society, these craftsmen relied upon their own labor and the health of the cloth trade for their subsistence. Befitting this social position, they argued a consistently independent mode of thought which called for a rough equality among all members of·society.

This intellectual independence first took a religious form that served to make the cloth-producing districts notorious centers of religious and social heresies.[13] The heretical voice of the clothworker has been recovered, in the person of William Bull, by A. G. Dickens in his examination of the court records of the Archbishop of York. It is at once an affirmation of the craftsmen's worth and a rejection of ecclesiastical hierarchy and the automatic deference expected by one's "betters." William Bull was a shearman (one of the most skilled and strenuous of the clothing occupations) who had just returned to his native Dewsbury from his journeyman's tramp through Ipswich and Hadleigh, both centers of cloth production and religious radicalism. One evening as he met with family and friends, the conversation turned to the parish priest (the year is 1543) and the Church itself. The resulting conversation attacked both doctrine and institution alike.

Bull reportedly considered his priest an unscrupulous "knave" to whom "he wolde not shewe his offenses," for if "he had japed [fornicated with] a fayre woman . . . the preste would be as redye within two or thre days after to use hir as he." Baptism he held up for particular scorn: "the [baptismal] fonte is but a stinking terne," he proclaimed, and last unction was nothing but "a sybertie sauce" that only interfered with his individual salvation. Indeed, he thought himself more likely to be saved "yf he had no suche sybertie sauce at his death." Bull finished what must have been a dramatic evening with a recitation of the Apostles' Creed, declaring in the end that "yf he belevyed stedfastlye in God, calling on God with a sory harte for his offenses, God would forgive him."[14]

The early history of popular Protestantism is written in Bull's soliloquy. So too is the antinomianism that would become the hallmark of radical sectarianism in the next century. Indeed, his anticlericalism and proto-Baptist beliefs were widely shared among artisans of the economically advanced regions of England, where Anabaptist and Familist tracts traveled clandestinely in the carts of London wool peddlers and itinerant cloth traders. Taken together these "heresies" formed what Christopher Hill has called a "layman's creed" that defined English popular religiosity for two hundred years before the advent of Methodism in the late eighteenth century.[15] With the emergence of this creed we find the first expression of a way of thinking which was unique (and notorious) among the craftsmen of early capitalist England. But if it was first cast in religious language, it did not long remain exclusively so. As early as the 1570s, when food shortages drove many rural clothworkers to the margins of subsistence, this popular religious radicalism quickly shaded over into the social activism of bread rioting as well as more clandestine and dangerous forms of social protest.[16]

This was one source of the small producer tradition: the artisan's personal and independent relationship with God coupled with an active devaluation of the clergy and the institutional church, both always available for service in more secular causes. The other source came from the journeymen and small masters of the London guilds. While the medieval guilds had been organized as mixed companies of merchants, masters, and working journeymen, by the middle of the fourteenth century all but the smallest of these guilds were controlled by wealthy export merchants who systematically undercut the position of the company journeymen and small masters. The result was the formation of a class of permanent wage-earning journeymen and the creation of independent journeymen's associations beginning in London at the close of the fourteenth century and spreading outward to Bristol, Exeter, and Plymouth during the next century and a half.[17] As these journeymen's associations contested for autonomy, higher wages, and representative rights within the larger guild organization, they began to articulate ideas that had lain fallow in the desuetude of earlier craft traditions and rituals. By the time the civil war overtook London it was, appropriately, an association of lesser tradesmen who transcribed the small producer ethic from journeymen's association to public remonstrance, pointedly reminding the "great men of England" that their "Flesh [was] that whereupon you Rich men live."[18]

These secular and religious ideas emerged from country cottage and journeyman's society into the light of public inspection during the early years of the English civil war. By the late 1640s, censorship of print and press had been eliminated and formerly suppressed popular ideas burst forth into the full light of day for the first time. It was in the midst of this flowering of hidden and suppressed ideas that the small producer tradition received its first serious public expression. This exposure came from a handful of London radicals who called the small shopkeepers and artisans of the City into action against Royalist England. The Levellers were a loosely knit group of radical thinkers and agitators who had originally come together in 1645 to defend freedom of religion against William Prynne's efforts to reconstruct a state church.[19] The Leveller leaders were themselves from dissenting backgrounds—John Lilburne was an orthodox Calvinist and an itinerant preacher, Richard Overton a General Baptist, and William Walwyn a staunch supporter of tolerance for separatists—and this religious radicalism provided the connection between them and the craftsmen who formed the backbone of London separatism.[20] Lilburne had, in fact, made himself the hero of London tradesmen by choosing the pain and humiliation of a public flogging rather than submitting to a judicial decree which ordered him to cease preaching his radical and popular religious views.[21]

The early religious concerns of the Levellers were soon overshadowed, however, by more secular concerns as their artisan followers began to make their grievances known.[22] What the artisans of London wanted was more than a new set of more liberal and tolerant rulers; they wanted true revolution. The focus of their grievances quickly fastened onto the contrast between the power and privileges of the wealthy and the poverty and powerlessness of working people such as themselves. "Look about you . . . see how pale and wan" the poor look; "how coldly, raggedly, and unwholesomely they are clothed," wrote William Walwyn, the Leveller leader closest to the minds of London working people. Then "observe the general plenty of all necessaries [and] the innumerable number of those who have more than sufficieth."[23]

The cause of all this misery and inequity was plain to see. And turning to a familiar artisan theme, Walwyn called to account men of large property who used "all means to augment their tithes and profits" at the expense of those who suffered as "the fruit of their labours [were] so unreasonably wrested from them [and] so super-fluously spent, or so covetously hoarded up."[24]

That Walwyn spoke for the whole of his artisan constituency was confirmed by their own petition to the Long Parliament which added an even sharper edge to his analysis. "Necessity dissolves all Laws and Government," the petition warned, "and Hunger will break through stone walls; Tender Mothers will sooner devour You, then the Fruit of their own Womb." The petition ended on an ominous note: "The Tears of the oppressed will wash away the foundations of [your] houses. Amen, Amen, so be it."[25]

What these London craftsmen were beginning to articulate was a formal criticism of the values and actions of the capitalist manufac-turers and merchant magnates who were denying the true value and independence of labor by subordinating artisans to them. In this criticism they offered an equation of their labor with the value of the commodities they produced and went on to claim a natural right to the full benefit of "the fruit of their labours." The tradesmen's peti-tion set out this idea graphically: "our Flesh is that whereupon you Richmen live, and wherewith you deck and adorn your selves."[26] "What are your ruffling Silks and Velvets, and your glittering Gold and Silver laces? [A]re they not the sweat of our brows, and the wants of our backs and bellies?"[27] To these men the revolution had become nothing more than a contention among the rulers over the division of the spoils: "Is not all the Controversie whose slaves the poor shall be?" they asked; "whether they shall be the King's Vassals, or the Presbyterians', or the Independent Faction's?"[28] They meant to put matters right.

The first step was taken by the Levellers in their draft constitution of 1649, the Agreement of the People. Arguing that Parliament ought to be the true *representative* of the people, the Leveller constitution provided for an equitably distributed franchise; the abolition of monarchy and the House of Lords; the election of sheriffs and justices of the peace; reform of the law to make it accessible and understandable to all, security of copyhold; the opening of enclosures; the abolition of tithes and the state church; the abolition of conscription, excise, and the privileges of peers; and the abolition of corporations and trading companies.[29]

These were the complaints of independent tradesmen and wage-earning journeymen against the restrictions placed upon them by large landholders and chartered merchant companies. The first five provisions sought to integrate this ascendent class into the political nation while eliminating the monopoly that property held over English politics. This notion, that all property holders—large and small—should participate in the governing of the nation, was a key point in the Leveller program and the small producer tradition as well.[30] Property—in tools and skills, as well as in land—gave its holder an interest in the nation, the Levellers thought, just as it conferred the responsibility to protect one's dependents and the deserving poor.

The last four points of the Leveller constitution were designed to maintain the economic viability of small producers. Security of copyhold and the opening of enclosures would insure that England would be a nation of small, yeoman farmers, the first by guaranteeing their title to their land, the second by preventing large accumulations of land in the hands of a few. The abolition of tithes and the state church would have protected the small farmer by removing the constant demands which the Church placed upon the fragile livings of many small producers, especially in bad times. And the abolition of excise and conscription would have freed farmers, tradesmen, and artisans alike from compulsory contributions, both of money and of sons, to what had become the English war machine.

The abolition of monopolies, in the form of chartered corporations and trading companies, was the key point in the Leveller's economic program.[31] Monopolies were crown warrants for small groups of merchants to engage in a specific trade without interfering competition. The Merchant Adventurers and the East India Company are the best known of these monopolies, but there were many others and they were the established way of doing business in England from late medieval times. The small producers who made the goods which the companies traded were dependents of these corporations, while itinerant provincial traders served as intermediaries, buying small lots of goods to sell to the London and Bristol companies and returning raw materials and some manufactures to local mechanics and tradesmen. The Leveller economic program was aimed at overturning this system and with it the grip which the merchant companies held over the lives of these petty producers and traders. The abolition of monopolies would have allowed them an independent status in the marketplace and would have indeed freed all national markets from monopolistic controls.

The theme running through and uniting these grievances was the same one that characterized the small-producer creed: the preservation of one's independence. We can follow this theme in thought and action from its spiritual basis in the layman's creed, through the demand for political inclusion, and on to the attempt to provide for an independent livelihood in an open market. The Leveller program gave voice to the needs and aspirations of an emerging class which saw as its antagonist an entrenched body of large landholders and merchant magnates. Had it been realized, the Leveller program would have removed the impediments which constantly threatened to reduce artisans, tradesmen, and farmers to a state of economic dependence on and political subordination to the will of the rich and powerful. [32]

The Putney debates of 1647, in which Cromwell foreclosed any possibility of popular suffrage, finished the Levellers as a constitutional party, however. Alarmed by the popular attempt to gain a measure of power in town and country, the vying factions in Parliament put aside their differences and closed ranks against what they saw (incorrectly) as a threat to property itself. As Edward Thompson has remarked, the response of Cromwell's son-in-law, Henry Ireton, to the arguments of the Levellers and army Agitators "seem like prescient apologetics for the compromise of 1688." [33]

Though the radical movement had failed in its infancy, the Levellers had gone far toward speaking for the working artisan and had articulated his grievances and aspirations with a sophistication never before attained. After the restoration of the monarchy in 1660, the small-producer tradition was forced to live an underground existence, returning again to the more personal debates of shop, small congregation, and neighborhood tavern. Its ideas about political rights and economic independence might surface again in the abortive Popish or Rye House Plots of 1678 and 1683, but as a social movement, London radicalism was dead.

The intellectual legacy of the civil war period was, however, manifold: it included a government answerable to the people; a franchise that included property in skill and labor, as well as in land and stock; and a hatred of monopoly and accumulated wealth which subverted the natural economy of small craftsmen. Perhaps the legacy was best expressed in a Leveller pamphlet written in the early years of retrenchment. In *More Light Shining in Buckinghamshire* artisan values received their sharpest edge before the nineteenth century.

> When a man hath got bread, viz. necessaries, by his labor, it is his bread; now the other that sweats not at all, yet makes this man to pay him tribute out of his labor . . . it is theft. Mark this you great Curmudgeons, you hang a man for stealing for his wants, when you yourselves have stole from your fellow brethren. So first go hang yourselves for your great thefts.[34]

We will have occasion to hear these thoughts expressed again, but under conditions of hopeful expectation rather than frustrated defeat.

III

The democratic aspirations of the common people in the English civil war found expression not only in politics but perhaps most profoundly in religion. The civil war, for all of its political import, took place in an acutely religious age. The passions of the Reformation had not yet cooled in the hearts of many English men and women when the release of countervailing authority during the

1640s and 1650s led to the formation of sect upon sect, each project-
ing in its own way God's design for humankind.[35] It was ultimately
in the theology and practice of these sects that the democratic possibilities foreclosed in the defeat of the popular movement were carried
on, both in thought and in action.[36]

One of the most significant of these sects, if it can properly be
called one, was the Ranters. The Ranters had neither organization
nor formal liturgy, but they had a powerful idea.[37] God, for the
Ranters, was immanent: He existed in "man and beast, fish and fowl,
and every green thing, from the highest cedar to the ivy on the
wall."[38] Moreover, the Ranter God was never divorced from human
concerns but existed in every person of whatever station or condition
of life. According to one early Ranter, the relationship between God
and the individual was simple and direct: "he is me and I am him."[39]

It was to this notion of an immanent God dwelling equally within
every soul that we can trace the doctrine of "inner light" that formed
the foundation of early Quaker beliefs.[40] Early Quakerism shared a
hazy connection with the civil war Ranters and incorporated many
Ranter ideas into their theology and organizational forms. It was, in
fact, only during the 1650s, when the unwelcome gaze of civil and
ecclesiastic authorities fell upon the sect's founders, George Fox and
James Nayler, that the Quakers began to differentiate themselves
decisively from the Ranter culture.

It was to the taming of the spiritual individualism inherent within
the doctrine of "inner light" that Fox and William Penn applied their
considerable energies. By the late 1650s, with the English revolution
in its death throes, official Quakerism sought at least enough accommodation with authority to ensure the survival of the sect. In his
preface to Fox's *Journal*, Penn attacked the antinomianism of the
Ranter wing, which he claimed "would have had every man independent, that as he had the principle in himself, he should only
stand and fall to that, and nobody else."[41] The path upon which Fox
and Penn set out was one which had some form of national church as
its logical terminus, and by the end of the 1670s official Quakerism
had been purged of its overtly individualist elements by the defection
of the Proud Quakers, whose appeal was decidedly plebeian, and by

the Story-Wilkinson separation of those who opposed Penn's and Fox's delimitation of the "inner light." After 1680, Quakerism remained radical, but only when viewed in the context of the Restoration rather than the revolutionary years.

It was in this mood of accommodative Quakerism that William Penn made his plans for the colony of Pennsylvania. Gary Nash has analyzed the pressures under which Penn drafted his successive frames of government and has concluded that Penn moved from the liberal "Fundamental Constitutions" to the oligarchical Frame of Government of 1682 in order to calm the fears and accommodate the desires of wealthy English investors.[42] Yet despite Penn's suppression of certain popular tendencies within Quakerism and his shift to a more authoritarian Frame of Government—Algernon Sidney called it "the basest laws in the world and not to be endured or lived under"—the Frame of 1682 contained much that would have pleased both the Levellers and their artisan constituency.[43]

The Frame of 1682 provided for nearly complete religious toleration, instituted a judicial system that was accessible to the common person, established a penal code purged of the current barbarities of the English and continental systems, and provided for secret ballots and rotation in office.[44] In all, Penn's final Frame was a constitution of accommodation. Just as Penn and Fox had trimmed the antinomian tendencies from the Quaker movement to accommodate it to a reconstituted monarchy, Penn's Frame of 1682 trimmed the authority of the popularly elected Assembly to accommodate the wishes of his investors. In Penn's eyes, Pennsylvania was to become a utopia; but it was a utopia of rather a distinctive type. The Frame of 1682 attempted to strike a constitutional balance between a political and administrative hierarchy and an assertive popular individualism. In a society of obedient and disciplined church members such a utopia might function well enough; in the all-too-real world of Pennsylvania it was doomed from the start.

Penn was by place and temperament an aristocrat. As such he saw no contradiction in the balance which he attempted to maintain between hierarchy and individual liberty in Pennsylvania. He chose not to attend to the counsels of those early Friends, purged in the rise of official Quakerism, who pointed out as they moved through Muggletonianism and the more esoteric sects to deism and, at times, outright atheism, that the doctrine of "inner light" was profoundly democratic and would in time come to stand alongside its political counterpart, the rights of the "freeborn Englishman." It was a lesson that Penn and his heirs were soon to learn.

It is from this vantage point that we can best understand Pennsylvania's long history of popular anti-authoritarianism. Almost from its inception in 1681, rival factions in the province had sought support from the laboring classes in their political battles with one another.[45] In doing so they drew upon discontent felt by artisans and other workingmen and gave it a legitimate and respectable public venue. This, in turn, made popular participation in provincial politics a recognized, if not easily accepted, tradition.[46]

As a proprietary colony, every aspect of life in Pennsylvania, from the sale of land to the administration of justice and the approval of provincial laws, was subject to the will of William Penn. He appointed the provincial governor and his councilors as well as the officers of the provincial courts. Moreover, through his ample provision of places in the new colonial government, Penn attempted to control the provincial Assembly with a generous system of patronage, title, and prestige.

These proprietary prerogatives, exercised from afar, rubbed sharply against the grain of Whig principles and the traditions of popular rights and Quaker anti-authoritarianism that informed Pennsylvania's early political culture. Pennsylvanians had been granted rights which were unique in the early modern world. There was no state church to demand fees and tithes, and religious toleration was nearly complete. The freehold franchise encompassed most adult males in a society in which property-holding was the rule rather than the exception. Moreover, the penal code was the least punitive of any contemporary state, while the judicial system was more democratic than any before the nineteenth century. Set against this uniquely liberal structure of government, the exercise of proprietary rights was bound to create friction when confronted with a free population three thousand miles away.[47]

And indeed it did. From the rejection of his Frame of 1682 by a General Assembly that thought it corrupt and overbearing, Penn fought a continuous battle with his colonists who were determined to forestall all proprietary incursions upon their newfound liberties. Thus only a decade after founding his colony, Penn found himself forced to return provincial government to local hands in the face of a general uproar over his appointment of John Blackwell, a Puritan, as Pennsylvania's governor. [48] Less than a decade after that, Penn was himself in Pennsylvania attempting to reassert his authority over the colony, but, in the end, he returned to England having been forced to grant a long list of concessions that strengthened not his, but the colonists' position. [49]

Penn's concessions created an opening in the web of proprietary authority large enough for important precedents to be established. In the first decade of the eighteenth century David Lloyd took full advantage of these concessions and laid the early groundwork for what would become a tradition of popular politics in Pennsylvania. [50] As speaker of the Assembly, Lloyd carried on a powerful and vituperative campaign against Penn's proprietary authority as it was embodied in the person of James Logan, Penn's provincial secretary. Against Logan's attempts to maintain proprietary control, Lloyd turned to the popular press and painted Logan in the colors of despotism, arbitrary authority, and oppression. While Logan, in turn, decried Lloyd's program as leading to popular tyranny and eventual anarchy, the public debate between them polarized political thought in the province and pushed it toward its limits. Whether he acted from high principle or simple expediency, Lloyd increased the range of public participation in provincial politics, advertising forthcoming elections, printing Assembly proceedings, and introducing election slates into everyday political practice. [51]

We do not know in any direct way the effect this controversy had on the minds of Philadelphia craftsmen, yet it is clear that it simplified choices among the city's laboring classes. In taking the first steps toward popular politics, Lloyd established a popular precedent that, at least initially, could only have added to the independence of mind of artisans and other workingmen. But, the evidence also reminds us, the people of Philadelphia were only invited into the political theater on condition that they sit, hands folded, and watch the spectacle performed before them. They were invited as auditors (taking up the cheaper seats, of course), but not yet as actors on the stage.

In the years after 1710 the controversy became a memory dimmed by better times and full employment. The Lloydians became a minority party and lost the Assembly speakership entirely. Wealth once again ruled the province uncontested.[52] But this quiescence was to be short-lived. The angry contentiousness of the Lloyd-Logan dispute had demonstrated both the waning of Quaker utopianism and the growing divisiveness of Philadelphia society. What had been a mildly differentiated community of Quaker immigrants in the 1680s had by the early eighteenth century become an increasingly class- and religiously divided city. Freed from the legal and economic liabilities suffered by religious dissenters in Restoration England, Quaker merchants had prospered in Philadelphia, although at a cost to the spiritual aspects of their faith.[53] By the 1720s, the egalitarian side of Quakerism found little resonance among the city's mercantile elite, who had long since turned their attention to the private pursuit of profit and power, abandoning Quaker communitarianism to the lesser artisans and farmers of the province.

While Mammon divided the Quaker community, the appearance of German, Irish, and Scots-Irish immigrants in Philadelphia after 1715 recomposed the city's laboring classes.[54] Fleeing poverty and forced conscription in their homelands, German artisans and Irish and Scots-Irish indentured servants flooded into Philadelphia, bringing their Lutheran, German Reformed, and Presbyterian confessions with them. The arrival of hundreds of these immigrants each year injected an element of religious and ethnic diversity into Quaker City life that quickly made Quakerism a minority faith in the province and created a solid base of religious pluralism among the city's laboring classes.[55] Set apart from the ruling Quaker elite by their social origins and position, and excluded from the Quaker artisan community because of their diverse religious affiliations, these immigrant craftsmen formed a fertile ground for political mobilization once a popular leader appeared. They would not have long to wait.

In 1722 Philadelphia fell into the throes of its first real depression. Caused by the collapse of the South Sea Bubble and exacerbated by a long-term decline in the city's West Indian trade, the depression brought the seaport's economy to a standstill. For the first time in the history of the prosperous colony, Philadelphians experienced falling prices for their products, the near cessation of shipbuilding, and an overall decay in the city's overseas trade. For the city's craftsmen, however, the worst effect of the depression was the general disappearance of specie and with it the disruption of everyday economic life. Without paper currency to use as a substitute for specie, farmers could not sell their produce, craftsmen their goods, nor journeymen receive their wages. Paper money thus became the electric issue of the 1720s, pitting the middling and lower ranks of the Quaker City against men such as James Logan, a merchant whose wealth not only insulated him from the depression but placed him in a position to profit from it. Together with a small faction of wealthy merchants in the Assembly, Logan opposed the printing of paper money because its tendency to depreciate weakened their position as the province's chief creditors and gave the colony's debtors what they saw as an unfair advantage. As Logan put it, with an insensitivity typical of eighteenth-century men of wealth, "Sobriety, Industry and Frugality" had made him rich and anyone who was not as comfortably situated need only look to their own neglect of these virtues for the cause.[56] Needless to say, his outlook was not shared by the majority of Philadelphians who, though they practiced these virtues assiduously, continued to find themselves poor, unemployed, and deeply in debt.

While we now attribute the causes of the depression of the 1720s to factors external to Philadelphia, the working men and women of Philadelphia who found themselves sliding fast into debt, poverty, and potential ruin took a more impassioned and personal view of their situation; after all, they were the ones who looked into empty pantries and languished in the city's debtor's prison. In their search for explanation and remedy, Philadelphia's laboring classes turned their gaze inward and focused on the disparity of condition between those few who prospered from the depression and the many who, like themselves, were poised on the brink of disaster. In the prosperity of the preceding decade ordinary Philadelphians had watched a rising group of merchants prosper from trade, land speculation, and the Iroquois fur trade, but there was little friction and resentment for they had experienced good times as well.[57] Now with the economy at a standstill and craftsmen struggling for employment and subsistence, they found that the merchant oligarchy would do nothing to alleviate their suffering. The currecy debates of the 1720s drew upon these grievances and forged them into a powerful criticism of the social, political, and economic domination of Philadelphia's ruling class.

The first signs of popular discontent appeared in the assembly elections of October 1722 when Philadelphia artisans joined the small farmers of Pennsylvania in voting down all but two of the Assembly conservatives who had refused to authorize the printing of paper money.[58] In the following year they completed the task by removing the two remaining hard money legislators, merchants Samuel Carpenter and Robert Jones.[59] In these elections, which Logan characterized as "very mobbish and carried by a Levelling spirit,"[60] Philadelphia's merchant oligarchy found that the disrupting experience of poverty and indebtedness had brought small craftsmen, journeymen, and casual laborers to echo the words of "Roger, the mechanic": "We are made of the same Flesh and Bones, and after the same manner with [the rich], so that our Sense and Feeling of Happiness and Misery, Justice and Injustice, good Fortune and ill Fortune, are much the same in Us all."[61]

When they were again invited into Philadelphia's political theater, city craftsmen found that the script had been changed: they now had small, but important, parts to play. The script was offered, oddly enough, by Penn's appointed governor, Sir William Keith, and it was aimed at those "butchers, tailors, blacksmiths, journeymen, apprentices, porters and carters," who had secured their election tickets at Keith's Leather Apron Club.[62] A Scottish baronet and an Anglican, Keith had spent much of his youth at the Jacobite court in

France and as a young man had narrowly escaped a charge of high treason for his Jacobite sympathies.[63] Appointed governor by the Penn family in 1717 to replace the incompetent Charles Gookin, he quickly earned a reputation as an evenhanded and competent administrator. But sometime in the early 1720s Keith turned against Penn and the proprietary elite and took sides with the artisans, shopkeepers, and middling merchants of Philadelphia who called for the printing of paper money. The reasons for his change of allegiance remain obscure, but Jacobites had appealed for popular support among the small producers of England since the late seventeenth century, often in language identical to Keith's.[64] Whatever his reasons for joining the popular party, Keith punctuated his apostasy by choosing an old Leveller theme for his opening speech to the 1722 Assembly. Echoing the thoughts of William Dell and Richard Overton, as well as the civil-war radicalism of Gerrard Winstanley, Keith reminded the Assembly of the plight of Philadelphia's laboring classes. "We all know it is neither the Great, the Rich, nor the Learned, that compose the Body of any People," he declared, "and that Civil Government ought carefully to protect the poor laborious and industrious Part of Mankind, in the Enjoyment of their just Rights, and equal Liberties and Privileges, with the rest of their Fellow Creatures."[65] Legislation derived "from these Principles," Keith went on to argue, would "very much tend to calm the Minds of the People."[66]

Keith's gesture toward the people was more than mere rhetoric. Until he left Pennsylvania in 1728 to repair his declining fortunes in England, Keith stood at the center of a heated and closely fought battle with Philadelphia's Quaker elite. Much more than the demagogy with which he was charged by his opponents, Keith's confrontation with Philadelphia's ruling class drew its form and energy from the artisans and shopkeepers who met at the Leather Apron Club. The club, also known as the Tiff Club in recognition of a popular laboring-class drink, was grafted onto that center of artisan conviviality—the tavern—and sought to forge connections between the developing political awareness of city craftsmen and Keith's antiproprietary party. We know little of the everyday operations of the club, other than political matters were its mainstay, but it is certain that its mechanic members formed the organized foundation for the street celebrations following Keith's election to the Assembly, as well as the personnel for the popular intimidation carried on against hard-money supporters in 1728 and 1729.[67]

Until Keith's departure and the collapse of the Keithians in the early 1730s, the popular, anti-proprietary party dominated provincial affairs. For his part, Keith labored diligently to represent the men of small property in the city. Joined in mid-decade by David Lloyd, who came out of political retirement to engage Logan once again, and by the anti-proprietary merchant Francis Rawle, the Keithians created the first political coalition to include small producers since the days of the Levellers.

What the Keithians espoused was nothing less than the program of the Levellers transformed by the passage of time and the trans-Atlantic crossing. As Lloyd put it, "a mean man of small interest, devoted to the faithful discharge of his Trust and Duty to the Government, may do more good to the State than a Richer or more Learned man, who by his ill Temper and aspiring Mind becomes an Opposer of the Constitution by which he should act."[68] There is more than an echo here of the Levellers and their campaign for the rights of small producers. In his vindication of Pennsylvania's popular government Lloyd reiterated the principles of popular representation argued by the radical Levellers at Putney who thought "the poorest he that is in England" had an equal claim to representation in government as "the greatest he."[69] And in his claim that a poor man might make a better legislator than a rich one, Lloyd moved closer to the social goal of the Levellers, the creation of a republic of small producers.

James Logan himself gave grudging testimony to the persistence of radical civil-war themes among Philadelphia's middle and lower ranks. In his *Antidote* to the efflorescence of popular thought and mobilization during the 1720s, Logan compared the political defeat of the proprietary elite to the demise of the monarchy in civil-war England. "Both Kings and Parliaments have err'd," he admitted, "and Mankind have openly censured the one or the other (and some both) for certain Proceedings in *England* towards the middle of the last century."[70] This guarded allusion to the overthrow of the Stuarts reminds us that the memory of the English civil war was a persistent feature of eighteenth-century popular culture, just as Logan's reluctance to speak of the revolution directly suggests the depth of the official suppression of its ideas.

Casting a worried yet disdainful eye toward the popular potential of the Keithian movement, Logan continued his Anglo-Pennsylvanian comparison with a warning to the new, popularly elected legislature. Referring to the Long Parliament, which had overthrown and executed the king, Logan argued that "all men condemn their excessive abuse of Power," and suggested that the more recent "abuse of power" by the artisans and shopkeepers of Philadelphia would, in turn, be equally condemned.[71] Logan's advice to the people of Philadelphia was to be "grateful and quiet," to return to their places, and to cease "endeavouring *to be better than very well*."[72]

The controversy of the 1720s was thus more than a debate over the merits of paper currency. Not far beneath the surface was a serious social and political challenge to the colony's men of wealth from Philadelphia's small property-holders. This antagonism grew in the course of the decade and became more pointed with each passing year, so that by 1729, Philadelphia's political rhetoric had begun to acquire the class-inflected voice that would distinguish it for more than a century. In that year, the Keithians again attacked the leaders of the proprietary party—James Logan, Andrew Hamilton, and Jeremiah Langhorne—for their continued resistance to the printing of paper money. In their anonymous pamphlet, *The Triumvirate of Pennsylvania*, the language of contention is sharp and clearly drawn: gone is the rich-versus-poor imagery of Keith's Jacobite populism and in its place is an assertion of the class character of political opposition. In the rhetoric of the later Keithians, the oligarchy is no longer the set of overbearing individuals whom the anti-proprietary party had pilloried throughout the 1720s; instead they are now collectively portrayed in class terms, as "the most tyrannical aristocracy in the world."[73] Likewise, the merchants portrayed in the pamphlet are made to condemn laboring people, not for their moral failings, but as "these Levellers" who, if armed with "this Monster of Paper Currency," would reduce the wealthy of the city "to the same Condition [as] Tradesmen, Handicrafts[men], and farmers."[74] Though not yet a full-blown language of class, the appearance of *The Triumvirate* nevertheless marks the emergence of an idiom of structured opposition that would be the hallmark of Philadelphia radicalism for the next hundred years.

The biting rhetoric of *The Triumvirate* bears testimony to the education that Philadelphia's artisans and tradesmen had received at the Leather Apron Club. If anything, Keith's absence from the city after 1728 revealed his influence to be a moderating one. Amidst rumors that armed men from the surrounding countryside were preparing to march on Philadelphia in support of paper money, a journeyman printer entered a small but important argument in a very long dispute. Recently settled in Philadelphia and introduced to Keith and the Leather Apron Club, Benjamin Franklin entered the list in support of paper money with *A Modest Enquiry into the Nature and Necessity of a Paper Currency.*[75] This brief pamphlet is noteworthy in the history of popular radicalism, less for its anti-specie arguments than its rehearsal of William Petty's idea of the labor theory of value. In it, Franklin added the intellectual weight of English political economy to the artisans' small producer tradition. Labor, he told Philadelphia craftsmen, was the only proper "Measure of Values," and "The Riches of a Country" were, accordingly, "to be valued by the Quantity of Labour its inhabitants are able to purchase."[76] Even trade, the foundation of Philadelphia's ruling oligarchy, was "nothing else but the Exchange of Labour for Labour."[77] Franklin's pamphlet, coming at the end of a period of popular agitation and organization, proved an important weapon to the people of Philadelphia in their fight against accumulated wealth and power. Before 1729 no one had questioned the mercantile claim that commerce was the linchpin of a prosperous community. With Franklin's pamphlet the idea was now broadcast that the producers of society were the true pillars of every community.

This idea, that everything of value came from those who fashioned goods from their thoughts and labor, was already a staple of the small-producer tradition. Eventually, in the late eighteenth and early nineteenth century, it would form the foundation for an attack upon overbearing wealth in all of its forms. In colonial Philadelphia, it served to reinforce the working artisan's feeling of self-worth and confirmed his true value to society.

This new-found laboring-class ethos is echoed in William Keith's alimentary parable of the fat man.[78] In it a fat man, whose swollen girth represents overgrown wealth, confronts a more slender laboring-class crowd. "Lord what a filthy Crowd is here! Pray good People, give Way a little!" exclaims the fat man with typical upper-class arrogance. It is, significantly, a humble weaver who brings him low: "A Plage confound you for an over-grown Sloven! and who in the De[vi]l's Name, I wonder, helps to make up the Crowd half so much as your Self? Don't you consider . . . that you take up more Room with that Carcass than any Five here? Bring your own Guts to a reasonable Compass (and be b[un]g'd) and then I'll engage we shall have Room enough for us all."[79] The images are homely, but Keith's message to Philadelphia's men of wealth was clear: begin to equalize wealth and resume "the Moderate Prudent Way of Quaker Oeconomy" or face the wrath of an awakened artisanry who will redistribute wealth by closing up (bunging) the greedy mouths of overbearing merchants and landowners.[80]

By mobilizing, organizing, and educating the working people of Philadelphia, the Leather Apron Club changed the face of popular protest and artisan radicalism in the Quaker City. Through their active participation in Keith's popular movement of the 1720s, Philadelphia artisans and workingmen began to cross the critical divide separating crowd action and coalition politics.[81] Although it has escaped the notice of most modern historians, the entry of Philadelphia's laboring classes into the political arena during the 1720s marked the beginning of America's first working class.

IV

Following Keith's departure in 1728 and David Lloyd's death in 1731 the popular anti-proprietary party found itself without its two most important leaders. Cast adrift, the ex-artisan shopkeepers and small merchants who dominated the Keithian movement made their peace with the proprietary party and left the leather aprons to fend for themselves in the political wilderness. This, and the return of prosperity in the 1730s, seemed to quiet the laboring-class ranks. But the lessons of the 1720s were there, well learned and ready for use.

The immediate legacy of the popular agitation of the 1720s is difficult to evaluate. Although the political waters appeared calm during the 1730s, there were ripples enough to suggest significant tensions below. From 1727 until the 1750s, Benjamin Franklin met with the members of his famous discussion circle, the Junto.[82] In its composition—three printers, a shoemaker, a joiner, a clerk, a surveyor, and a young gentleman—this circle approximated the membership of Keith's Leather Apron Club, as did the half a dozen other clubs that formed in imitation of the Junto. The fact that contemporaries routinely referred to these clubmen as "the leather aprons" suggests a strong continuity with the organization of the 1720s.[83]

More concretely, the appearance of John Webbe's "Z" letters in the *Pennsylvania Gazette* confirms the persistence of popular radicalism even in flush times. A lawyer and part-time editor, Webbe aimed his barbs at the men of wealth who again ruled the Quaker City.[84] In his opening essay he compared Philadelphia's social structure to the architecture of a building, warning that "if the Superstructure is too heavy for the Foundation, the Building totters."[85] He went on to draw a further parallel between the struggles of the English people against the Stuart monarchy in the seventeenth century and the ordinary Philadelphian's continuing struggle with the city's Quaker and Anglican oligarchy. "During the inglorious reigns of the *Stewarts*," Webbe declaimed, "it was a perpetual struggle between them and the People, those endeavouring to subvert, and *these* bravely opposing the Subverters of Liberty."[86] The outcome, he reminded the oligarchs, was that "one lost his life on the Scaffold, another was banished, [and] the Memory of all of them stinks in the Nostrils of every true lover of his Country."[87]

Webbe was more than an iconoclast, however. In a declaration of principle remarkably like that offered by the Levellers at Putney in 1647, Webbe declared that "Freedom is the Birth-right of every Man."[88] "We are all born equal," he continued, and it would be "Impudence to tell another Animal like myself that I came into the World his superior; none is born with a Right to control another."[89] Here were the themes of four centuries of English popular-radical thought collected and held open to public view. Then, with an argument that traveled beyond the Levellers toward the Diggers and the materialism of the laboring-class religious sects, Webbe asserted that "Nature has made no distinction; from the same Clod of Dirt she forms a Monarch and a Cobbler."[90] The world, Webbe declared, was "one Great Commonwealth and its inhabitants [were its] Fellow Citizens."[91]

From the Leather Apron Club of the 1720s and the quiet agitation of the 1730s it is a short journey to Franklin's militia Associators of 1747 and 1755. Modeled after Cromwell's New Model Army, these voluntary militia companies drew wide support from Philadelphia's working classes; and it was from this base that Franklin built a new anti-proprietary party that made Philadelphia artisans a "strong and even formidible" force in city and provincial affairs.[92] We should not be misled by the military trappings of Franklin's militias; they were, in fact, more social and political clubs than true military units. But they provided a vital forum for the discussion of political issues at a time when few other avenues were available. And in the process of meeting together, Philadelphia artisans honed their organizational skills, elected their own officers, and made important connections across class lines.[93]

The script was now finished and the laboring classes found in it a small part for themselves. To the Assembly factions their parts may have been minor, but to craftsmen, journeymen, and laborers it was a challenging and important new role. And it was possible, given the shock of new experiences, that the habit of discussing and criticizing government policies, and the drawing of class lines, could prove contagious.

The Seven Years' War was soon to provide many such experiences to Philadelphia's laboring classes. Britain's imperial wars had cast their specter across the North American colonies since the late seventeenth century, but none was to have the impact upon Philadelphians as did the Anglo-French conflict of 1756–63. Before 1756 Pennsylvania had been relatively isolated from the effects of war because of its geographical location and the pacifism of its Quaker-dominated Assembly. Now Philadelphia was made one of the central provisioning centers for the British army and navy. As Philadelphians soon discovered, the social effects of a wartime economy were enormous.

In the early years of the war privateering and military contracts changed the structure of opportunity in the Quaker City, offering unprecedented wages to workingmen and immense profits to the city's merchants. But as quickly as the war created these overheated conditions, the shift in the British theater of operations to the West Indies in 1760 rapidly cooled Philadelphia's new-found prosperity. After 1760 Philadelphia fell into a precipitous economic depression, the bounds of which few had ever imagined possible. As always, workingmen were the first to experience the disruption. Employment, which had been so plentiful a few months before, dried up with the termination of war contracts and the removal of the British army and navy. And compounding the burden of sudden unemployment, the war had also filled the city with hundreds of jobless soldiers and sailors who competed with local craftsmen in the work-starved economy.[94]

Despite these problems, the disruption of the city's economy might have been more tolerable had prices fallen along with the prospects for employment. But the interplay of continued wartime demand for foodstuffs, European grain shortages, and local crop failures kept prices high and drove growing numbers of laboring-class families to seek poor relief. This insecurity of subsistence attacked the artisan's pocketbook, but also the mainstay of his value system: his competency to provide for this family's needs through his labor.

Declining wages, diminished opportunities, growing unemployment, and high prices made it difficult for even middling craftsmen to maintain a respectable standard of living; for the laboring poor it meant that family subsistence required public or private aid.[95]

Set against the sense of frustration and unfairness felt by craftsmen and their families was the equally disconcerting experience of an undisguised display of wealth in the city. The early wartime boom that had brought full employment to the working classes had also brought what can only be described as super-profits to the city's merchants and employers. But while employment declined precipitously after 1760, prices, and with them merchants' profits, continued to rise, leaving Philadelphia an even more polarized community. Using modern statistical techniques, we can trace the changes in social structure wrought by the Seven Years' War with considerable accuracy: community wealth clustered at the top as it always had, only in greater degree, but now the middling group of respectable artisans and shopkeepers who had figured so prominently in the past history of the city controlled an ever-shrinking portion of the city's wealth.[96]

Laboring-class Philadelphians, of course, did not experience the increasing maldistribution of wealth in exactly this way. Around them they saw men like themselves scrambling for work, accepting wages below custom, working outside their trades, and sacrificing their dignity to feed their families. And as they looked beyond their own ranks, they found a new world of imported coaches, liveried servants, and urban mansions built in the latest Georgian style. The threat to their competency that the postwar depression had brought in its wake combined with this arrogant display of wealth to bring craftsmen, journeymen, laborers, and petty tradesmen together to discuss and evaluate the new direction their community was taking. These discussions would soon lead them to question the wealth and authority of men who built mansions from the profits of war and to question the imperial system that had manufactured a war which caused so much distress. Their aim, beyond the restoration of a functioning economy, was best expressed by the Committee of Privates in 1776: "We are contending for the Liberty which God has made our Birthright: All Men are entitled to it, and no set of Men have a right to anything higher."[97]

There is a hazy dialectic that operates between patrician and plebeian when the former falters and the latter has learned something of the mechanics of power. Such a dialectic operated in Philadelphia in the years following the Seven Years' War. It began when Benjamin Franklin, who had spent much of his middle age organizing an artisan following for his anti-proprietary party, sought to extricate the party from its impasse with the unpopular proprietor, Thomas Penn, by negotiating a royal charter for Pennsylvania.[98] That he did so in expectation of finally eliminating the power of the avaricious, militantly anti-Quaker Penn was of little importance; that he did so as the British ministry was in the process of reorganizing the colonial relationship between England and America was to have a profound impact. By his intimate association with the very forces who were then drafting the Stamp Act, Franklin drove a wedge into the very laboring-class cohesiveness he had worked so long to achieve. Franklin's action, and those of his anti-proprietary party, sent a shock of confusion through the laboring-class ranks of Philadelphia. At no time was the change in Franklin's standing more dramatically illustrated than in September 1765, when Franklin's political partner, Joseph Galloway, had to deploy 800 loyal artisans to protect Franklin's property from the wrath of an angry crowd which sought to punish the man they thought responsible for the Stamp Act.[99] To many working people, Franklin's blunder was a betrayal of the principles they had rallied to in their clubs and military associations; to many more it was a very personal betrayal of the bonds of respect, affection, and esteem which had bound Franklin and city craftsmen together.

A less personal but more instructive lesson came in 1769 when Philadelphia's merchant community turned its collective back on the city's popular non-importation movement. The last major group of American merchants to adopt non-importation, Philadelphia merchants were among the first to resume normal trade with Britain in 1770. Sensing that their Whiggish cloak of moral legitimacy was wearing dangerously thin, Philadelphia merchants met popular criticism of their dissimulation with an air of defensive haughtiness, declaring that city artisans were " a Rabble [who had] no Right to give their sentiments respecting an importation."[100] Philadelphia's artisans took this notice for what it was: a sign of moral weakness and a lack of resolve on the part of the city's "natural aristocracy." The merchants had violated the first canon of their own Whig thought: they had elevated their own particular and selfish interests above the interests of the community and the common good. In the eyes of the working community, the merchants had also made a tacit alliance with "old corruption" by resuming trade with England and, in so doing, had only confirmed their moral unfitness to rule.

The years from 1764 to 1770 were thus years of confusion for the laboring community. In these few years they had seen their champion, Franklin, consort with corrupt English ministers and placemen in what appeared to be an attempt to kill proprietary government at the cost of extending the hand of English corruption and tyranny over their own province. Then they had watched the same merchant class in whom they had placed their trust and suffrages, and from whom they had learned about the dangers of corruption and unbridled private interest, stumble down that very path.

Confusion, however, is the workshop of new ideas. One cooper remembered an old play his grandfather had attended, but couldn't remember its name. A cordwainer rummaged through the musty library of culture and found a fragment of a title. In time, the scripts of that old dramatist, Sir William Keith, now all brittle and yellowed with age, were taken from the shelf, dusted off, and read again. When they had finished reading and remembering, the working community could begin to rebuild what had been lost in the years of confusion.

What they had forgotten in those years came back to them with a shock of recognition: the responsibility for a just community lay in their own hands. The old Christian idea had, of course, been there all along, prodding their consciences: "The poor are those in whom the blessing lies, for they first *receive the gospel*."[101] The words are Winstanley's but they had a much longer lineage. More secular ideas had also been there, adding their own weight: "The poorer and meaner of this kingdom, have been the means of the preservation of this kingdom," claimed one agitator at Putney.[102] And, more recently, Sir William Keith had reminded the provincial Assembly that "it is neither the Great, the Rich, nor the Learned, that compose the Body of the People."[103] These religious, social, and political ideas mixed, interrogated each other, argued, and drew up a public declaration: henceforth the artisans of Philadelphia would think and act for themselves.

There followed, after 1770, a remarkable flowering of political activity on the part of city mechanics. In May 1770, artisans organized the first Mechanics' Committee, which delayed, if it could not ultimately halt, the resumption of trade with Britain. By 1772, mechanics were opening up the mysteries of civil government, calling for the publishing of Assembly debates and roll calls in the popular press and the opening of public galleries in the legislature. A few months later, they were demanding economic reforms that would benefit the laboring community as a whole.[104]

The turn toward separation and, later, revolution was itself first taken by radical mechanics and workingmen. By 1774 their organized strength enabled Philadelphia artisans to dislodge conservatives from the city's Committee of Inspection and Observation and to replace them with men of their own class.[105] In the final four Committee elections preceding the Declaration of Independence the number of artisan committeemen increased steadily until, in early 1776, they occupied nearly half of the total seats.[106]

The most radical of the popular organizations to appear in this period, however, was the Committee of Privates, a laboring-class militia company remarkably similar in design and function to the Agitators of the New Model Army. The historian of the Philadelphia militia has been able to account for fourteen members of the Committee: two schoolmasters, a professor, a shopkeeper, a clerk, a bookseller, a merchant, a distiller, a tailor, a cordwainer, and a carpenter—all men of modest means.[107] Like their seventeenth-century counterparts, the Committee of Privates provided a political education for its members at the same time that it helped to organize popular support for the Revolution. The critical educative role of the Committee is confirmed by the occupations of its leaders—two schoolmasters (William Thorne and Thomas Nevil) and a professor (James Cannon)—and by Cannon himself, who wrote that the members of the Committee had "been constantly enlarging our views, and sketching them beyond the first bounds."[108]

The purpose of this education was to equip those rank-and-file Associators "whose public spirit far exceed their abilities in the point of Fortune" with the knowledge and organizational skills necessary to take an active part in the governance and defense of the province.[109] In working toward this goal, the Committee of Privates spoke in a distinctly laboring-class voice; their calls for unrestricted manhood suffrage, the election of militia officers, long terms of enlistment, and public works projects to alleviate unemployment, and even their demand that "the people [have an] equal consultation" in the design of militia uniforms so that they were "within the compass of almost *every* person's ability," were aimed at the concerns of the working artisan and his community.[110]

Taken together, the activities of these radical-popular committees led to a change of mood within Philadelphia's laboring-class ranks; cut loose from their ties to moderate merchants and organized in their own associations, working people began a process of self-definition that would eventuate in their becoming a working class. This change in attitude can be measured in the 1776 "Memorial of the Committee of Privates," in which the militiamen informed the Continental Congress that they were "under the sense of being oppressed by the very men whose liberties and estates they are called out to defend."[111] Or measured again, this time against engrossers: "[merchants are] monsters unworthy of the fellowship of society . . . for as the monsters of the sea devour the small, so these endeavor to oppress and destroy the poor."[112] And, in a gesture recalling English practice, "Luke, the Physician" called for the seizure of engrossed goods to be sold by "gentlemen of character . . . at a reasonable price to all men alike."[113] Merchants are here no longer educated and capable partners in a common cause; from non-importation onward they were increasingly seen by laboring people as a class whose interests ran at cross-purposes to their own. It is a very short distance from here to the journeymen's benevolent associations and the early trade unions of Philadelphia which formed late in the Revolutionary era.

CHAPTER 2

Property Rights and Community Rights:
The Politics of Popular Revolution,
1765–1779

In the half-century before the Revolution Philadelphia artisans had moved intermittently toward a rudimentary sense of class and collective interest through their participation in the radical-popular politics of the era. Now as they joined the Revolution itself, the British threat combined with their collective political experience and small producer values to give them a heightened feeling of mutuality as well as a consciousness of larger purpose. For city artisans in the summer of 1776, the Revolution promised to do more than throw off the burdens of an ever-encroaching British empire. It also promised to humble the overgrown class of merchant oligarchs who had dominated Quaker City life since the 1720s and to restore the competency and independence that had been slowly disappearing from artisan life. The Revolution thus meant more than national independence for city workmen; it pointed equally toward a time when once again "all ranks and conditions would come in for their just share of the wealth."[1]

In the event, Philadelphia artisans would find a mixture of success and disappointment during the Revolutionary years. Success came quickly with the adoption of a new state constitution that enfranchised virtually every white adult male in Pennsylvania and provided for the most radically democratic government in the infant nation. But beyond the flush of initial victory, the majority of workingmen experienced increasingly hard times, inequitable militia service, the

British occupation of their city, and the rupture of their alliance with the city's radical middle class. By 1779 the prospect for a republic of labor, so vital only three years before, appeared more distant than ever.

I

In 1776, laboring-class Philadelphians faced a rapidly changing economic and political world. Until the end of the Seven Years' War craftsmen and other workmen had found the Quaker City to be a prosperous place to ply their trades. Population had expanded rapidly in the middle decades of the eighteenth century, growing from slightly less than 5,000 inhabitants in 1720 to more than 10,000 twenty years later.[2] By the end of the colonial era, Philadelphia's population hovered near 32,000, making it the most populous city in North America.[3]

Trade, the lifeblood of the city's economy throughout the seventeenth and eighteenth centuries, suffered periodic fluctuations but on the whole increased dramatically after 1740.[4] As early as the late 1730s, Philadelphia merchants had cleared more tonnage from their wharves than their competitors in New York and shortly thereafter they surpassed even Boston, long the center of colonial maritime trade.[5] Reflecting this generally prosperous economy, the wages of craftsmen and laborers had been sufficient, before 1760, to insure a good diet, adequate housing, and, for the more skilled tradesmen, a moderate level of domestic comfort.[6]

This long-term prosperity, and the expectation of a lifelong competency that had accompanied it, disappeared in the economic downturn that followed the Seven Years' War. After 1760, Philadelphia artisans suffered through a prolonged depression that saw the gains of the past thirty-five years quickly wiped away. Unemployment, which every craftsman had experienced as a result of illness or poor weather, now took on chronic proportions, while wages, which had been unusually high during the war, dropped sharply in the early 1760s and again after 1770, driving growing numbers of working people closer to the margins of subsistence than anyone had thought possible. Adding to the general sense of distress, poor harvests and overseas demand kept food prices high, so that a craftsmen's earnings purchased poorer food in smaller quantities than at any time in the past.[7] Food prices remained so high, in fact, that in the early 1770s the Overseers of the Poor took the unprecedented step of distributing bread among the city's working poor.[8]

Thus in the decade preceding the Revolution the people of Philadelphia experienced truly alarming changes in the substance of their lives. Once the capital of "the best poor man's country," Philadelphia craftsmen now appeared to be witnessing the collapse of their small producer world.

There is perhaps no better way of capturing the economic turbulence of these uncertain times than by observing the changing structure of wealth within the laboring community itself. At the turn of the eighteenth century, middling craftsmen such as carpenters, coopers, and shipwrights claimed 17 percent of Quaker City wealth.[9] The depression of the 1720s did little to alter this pattern of wealth-holding: in the decade beginning in 1726, these middling artisans continued to hold 12 percent of the city's recorded wealth.[10] But in the decade preceding the Revolution, their share of Philadelphia's resources shrank to an alarming 5 percent, while lesser artisans and unskilled laborers together amassed no more than 1 percent of the city's collective wealth.[11] Over the same period, by comparison, the city's most prosperous men had increased their share of city wealth from a modest 25 percent at the beginning of the century to 56 percent on the eve of the Revolution.[12]

Taken together, these figures tell us much about the experiences of Philadelphia workingmen. On the one hand, they reveal the rise of a ruling class made up of merchants, lawyers, and proprietary officials whose wealth had grown rapidly after 1730.[13] These were the first families of Pennsylvania, who aped the styles, manners, and attitudes of the English gentry and who had spent the middle decades of the eighteenth century attempting to bring the whole of Pennsylvania society under their control. But while these magnates provided the city's laboring classes with a visible and antagonistic class against which to define their own social identity, more important was the shrinking share of wealth claimed by the city's working people.

These were men who had realized, or expected to realize, the crafts-man's ideal of political and economic self-sufficiency, men who espoused the tenets of the small producer tradition. It was men such as these who entered the Revolutionary era with their fortunes falling and their way of life under steady attack. Men in similar circum-stances had rallied to the Levellers during the English civil war and would rally to the Sans-Culotte cause during the French Revolution; in Philadelphia they formed the backbone of the Revolutionary movement.[14]

Woven into this pattern of declining fortunes and threatened com-petency was a final factor that rent the finely knit social fabric of the Quaker City even further. As trade contracted and prosperity re-ceded, growing numbers of craftsmen found it impossible to main-tain the bound laborers who had assisted them in their shops and yards.[15] From the early days of the colony, Philadelphia craftsmen had employed indentured servants and African slaves to perform the heavy and repetitious work that was required in every trade. Until the middle of the eighteenth century, in fact, servants and slaves had made up nearly 40 percent of the city's work force. But following the Seven Years' War, the ranks of bound laborers shrank dramatically, until by 1775 only about one in ten laboring-class Philadelphians remained unfree.[16]

The formation of a substantial wage labor force after 1763 created a new world for Philadelphia artisans. As the relationship between master and apprentice turned increasingly on the pivot of cash pay-ment, the old craft system, with its regular advancement from ap-prentice to master, began to give way to a more calculated and exploitative relationship.[17] After 1750 fathers who negotiated ap-prenticeship contracts for their sons no longer relied upon custom—a master's gift of tools, clothing, or money as the apprenticeship came to an end—but cast their contracts in precise, hard-cash terms.[18]

As master craftsmen gradually shed traditional craft respon-sibilities and faced their apprentices and journeymen as employers rather than fellow artisans, workshop relationships began to show signs of strain. No longer tied by the bonds of craft mutuality, jour-neymen increasingly moved away from their masters' households and provided for their own room and board.[19] Finding their masters overbearing and exploitative, they changed jobs frequently.[20] And, in the last quarter of the century, they formed craft societies to meet the impositions of masters who saw in their journeymen little more than abstract and interchangeable units of labor.[21]

Thus by the time of the Revolution, Philadelphia craftsmen had more than political and imperial issues to digest. In a decade and a half the city had begun to change from a place where craft traditions predominated, and a young craftsman could look forward to a life of useful toil and moderate compensation, to a new world of capitalist relationships where little was assured and even the provision of life's necessities was left to chance. The changing nature of work and the economy thus added a measure of anxiety to the debates over independence and brought many workingmen to question the nature of the society that the Revolution was beginning to bring into being.

II

Some measure of the society which Philadelphia artisans hoped to create in the early years of the Revolution can be found in the provisions of Pennsylvania's Revolutionary constitution. The making of the constitution was itself one of the great democratic events of the Revolution.[22] Faced with an intransigent Assembly that steadfastly refused to relinquish hope, however remote, for a reconciliation with Britain, the Philadelphia Committee of Inspection and Observation, half of whose members were local artisans, called a town meeting in the State House Yard on May 20, 1776. The meeting turned out to be one of the city's largest, attracting more than 4,000 people to the small square on Walnut Street. The open-air assembly quickly endorsed the Committee's resolution that independence was essential to the preservation of their community, the colonies, and indeed for the protection of liberty itself. And since the Quaker-controlled Assembly stood in the path of independence, the committee continued, the people had little alternative but to return government into their own hands and redraft the frame of government to serve their collective needs. Thus was a warrant given to the

Provincial Conference to call a constitutional convention with the purpose of drafting a new constitution for the Commonwealth of Pennsylvania.

In their call for a convention, the radical leaders of the Provincial Conference—Thomas Young, Benjamin Rush, Christopher Marshall, and Timothy Matlack—endorsed a democratic demand made jointly by the Committee of Privates and the German Associators of the city militia which called for the enfranchisement of all Associators of twenty-one years who had lived in the state for one year and had paid provincial taxes.[23] This provision, along with another which disenfranchised confirmed Tories and those who publicly opposed independence, opened wide the gates protecting the corridors of power. Craftsmen, journeymen mechanics, and laborers were now poised to choose and instruct their own representatives in a democratic election. This is surely one of the triumphs of the eighteenth-century popular movement; in 1776 the Committee of Privates at last realized the demands of the Agitators in the English civil war: suffrage for independent men whose property often amounted to little more than their tools and skills.[24]

The constitution that emerged from the convention in the fall of 1776 was in many respects a popular-radical response to the operations of provincial government from the time of Penn's charter.[25] Viewed from this perspective, its principal innovations were its democratic suffrage requirements and its division of executive power among a five-member Supreme Executive Council. The suffrage requirements simply acknowledged the Associators' demands for political incorporation made at the calling of the convention, while the fragmentation of executive power reflected Pennsylvania's long history of popular protest against proprietary rule.

But, while important, the provisions of the revolutionary constitution went well beyond these immediate popular concerns to articulate a democratic form of government that directly challenged the foundations of republican constitutional thought.[26] Steeped in the history of ancient and renaissance republics—all of which had seen popular citizenship fall to the tyranny of dictators and despots—classical republicanism mistrusted power and feared unchecked pop-

ular power most of all. By placing its faith in the common people of
Pennsylvania and structuring government accordingly, the revolutionary constitution cast off from the moorings of classical political
thought and charted a singular course toward true popular government. As one early democrat put it, "the reasoning from the constitutions of Rome and Athens to that of Pennsylvania is vague and
undetermined. They understood not . . . how to constitute [popu-
lar] assemblies to answer the purposes of good government."[27] The
drafters of the 1776 constitution could not have agreed more.

As a democratic charter, the Revolutionary constitution admitted
the existence of but a single interest in the state, that of the people.
Accordingly, the Constitution provided for a unicameral legislature,
based representation on the state's taxable inhabitants, extended the
franchise to the adult sons of taxpayers, gave to the Assembly the bulk
of the powers of government, and established a Council of Censors
to oversee the operations of the government as a whole. To maintain
an informed citizenry and to keep power within the reach of ordinary
Pennsylvanians, the Constitution limited the tenure of officeholders
to no more than four years in every seven, opened Assembly sessions
to the public, ordered the weekly publication of all roll-call votes,
and insured a dialogue between citizen and representative by delay-
ing final passage of all pending legislation until individual bills could
be published in the state's newspapers and debated by the public at
large.[28]

In its democratic radicalism Pennsylvania's Revolutionary constitution encompassed the demands put forward by the Levellers in
their Agreement of the People and anticipated the demands English
Chartists would make in the People's Charter 200 years later. The
radicals and workingmen of Philadelphia had achieved in 1776—
during the gestation of the American working class—what the British
working class would still be striving for in its early maturity. There
were problems to be sure, not the least of which was the fact that in
enfranchising tens of thousands of Pennsylvania farmers, the constitution had created a powerful rural interest that would often be at
odds with the interests of urban workingmen. Yet despite the ever-
present potential of rural-urban friction, the democratic constitution

of 1776 represented a real victory for laboring-class Philadelphians, who would count active political participation among their new-found Revolutionary rights.

But along with its achievements, the provisions and tone of the 1776 Constitution also reflected the tensions inherent within Philadelphia's popular-radical movement. At the root of these tensions lay the issue of property; for as subsequent events would make clear, not all Philadelphians viewed the rights of property in exactly the same way.

These tensions were already apparent in the Constitution's formal "Declaration of Rights." On the one hand, the Declaration followed Lockean precedent and guaranteed to all Pennsylvanians their rights to "Acquiring, Possessing and Protecting Property."[29] But, at the same time, it made the potentially limiting claim that government existed for the "*Common* Benefit, Protection and Security of the People, Nation or Community, and not for the *particular* Emolument or advantage of any Single man, family or set of Men, who are only part of that Community."[30] This, along with the narrowly rejected sixteenth article, which warned that "an enormous Proportion of Property vested in a few Individuals is dangerous to the Rights, and destructive of the Common Happiness of Mankind," challenged the whole notion of absolute property rights and affirmed the ultimately social basis of property.[31] This tension between the rights of property and the rights of the community lay dormant in the mixture of classes and ideologies that made up Philadelphia's radical-popular movement. It would require the pressure of Revolutionary events to precipitate the elements of that mixture into its constituent parts.

III

Small masters and independent artisans, as well as journeymen and other wage-earning craftsmen, were swept up together in the gusting winds of resistance and rebellion against British rule. From the end of the Seven Years' War workingmen had responded to the tightening of imperial bonds in common cause with the lawyers, doctors, teachers, and small merchants who were Philadelphia's radical leaders. In this process of political mobilization they had learned to match their own small producer principles against the radicals' Whig canon and could begin to see in the arrogance of wealth—Georgian mansions, liveried servants, and ornate coaches-of-four—evidence not only of "corruption" but of their own livelihoods and way of life sacrificed to the private accumulation of wealth and personal advantage. Yet, for all its inventiveness, the intellectual world of 1776 remained, in important ways, a middle-class English world. The ideas of the radicals, as embodied in the new constitution, were largely the working out, on colonial soil, of issues raised during the English civil war. There men of large property and privilege had united in the face of popular assertion and denied history its first small-holders' republic.[32] In Philadelphia, where the city magnates had abdicated their positions of power rather than support war and independence, men of property could only grumble that "power had fallen into low hands,"[33] and nod in agreement with the empty condescension of Quaker schoolteacher Robert Proud: "Of all the plagues, that scourge the human race, / None can be worse, than *upstarts*, when in place."[34]

The radical-popular movement had entered a breach in the wall of authority that was created, ironically enough, by the intransigence of the provincial ruling class itself. And as political power and consciousness increased among the city's artisans, they began to sketch out a new political lexicon and a revised grammar of government, a grammar that, one conservative writer feared, would in its *"damned simplicity"* make of the "higher classes . . . simple freemen."[35]

From this unstable mixture of embryonic class consciousness, middle-class Whiggism, and political dissent the pressure of Revolutionary events worked to push positions to their extremes, to separate people and interests, and to emphasize tensions which had, in quieter times, been only potentialities. The first sign of tension came as the Pennsylvania militia mobilized in the winter of 1776–77.[36] As the laboring class Associators looked down the assembled ranks, remade acquaintances, and shared their anxieties about battle, they also began to notice the unbalanced composition of their companies. Absent were most of the city's Quakers—including many of the city's principal merchants—who had bought their way out of militia service under the pretense of pacifism. Absent, too, were members of the upper and middle classes who, like the Quakers, had paid modest fines for non-attendance, or had hired cheap substitutes from among the city's laboring poor, and were comfortably, and profitably, pursuing their lives at home. This set them on edge, but no less so than their year-long struggle to secure adequate pay and to insure the subsistence and well-being of their families while they served in the field. Philadelphia's militiamen had, in fact, mustered only after they had received concrete assurances from the Committee of Safety that their grievances would be addressed and that their families would not suffer in their absence.[37]

There was, then, uneasiness within the ranks of the Associators, although an uneasiness as yet considerably diffuse and individual in form: grumbling, disaffection, and desertion remained its major symptoms. This was to change in January 1777. At the beginning of the month, the Associators experienced serious combat for the first time, enduring heavy British musketfire and lethal torrents of grapeshot before finally driving the British into retreat at the Battle of Princeton.[38] It was a sobering experience for those unused to the horrors of violent death and dismemberment, and the sights and sounds of the battlefield brought home the very real risks and sacrifices they had undertaken. It was a serious and determined group of men who returned to Philadelphia in late January after completing their tours with Washington's army. What they found there gave shape to their uneasiness and provided it with a name.

The Associators had complained since the summer of 1776 that militia obligations had been fulfilled with little enthusiasm by Quakers and the upper classes in general; they now found that some of these same shirkers had profited handsomely in their absence. As the war progressed, the merchants who had remained in Philadelphia rediscovered the lesson of the Seven Years' War: patriotism could become a sure and easy road to profit. As in the last of the colonial wars, military contracts lined the pockets of those with money and political connections while they allowed many middling merchants and subcontractors to join the ranks of the tolerably rich.

Moreover, the widespread scarcity and engrossing of goods drove prices upward in an unremitting spiral of inflation which only multiplied the capital of the rich while it cut ever more deeply into the pantries of laboring-class households. The result was that many Associators "found their business and customers . . . so deranged on their return, and engrossed by those who stayed at home" that they had little prospect for ever continuing in their trades.[39]

The list of grievances that Philadelphia Associators compiled on their return in 1777 was copied from an older list that had circulated since the time of the Seven Years' War. Then, as now, many merchants had made their way to wealth, less from honorable and community-serving merchanting than from dishonorable and self-serving engrossing and profiteering. Even William Williams, a Connecticut delegate to the Continental Congress, laid the blame for the "alarmingly extravagant" rise in prices on Philadelphia merchants for whom, he noted, their own gain appeared to be their "Summmum Bonum."[40]

The mood of the artisans and workingmen who together made up the rank and file of the militia was less polite. They reminded the community that while defending their country they had left their "families at every risque of distress and hardships, and at the mercy of the disaffected, Inimical, or self Interested, [indeed] the most Obnoxious part of the Community."[41] They went on to point out that monopolizing merchants, whom they described as their "Innate and Worse of Enemies," were "Avariciously intent on Amassing Wealth by the Destruction of the more virtuous part of the Community."[42] The final result of this monopolizing, the Associators declared, was that "every Article of Life or Convenience was rais'd upon us, Eight, ten, or twelve fold at least," leaving many "at a loss to this day what Course or Station of Life to adopt to Support ourselves and Families."[43] An anonymous writer to the *Pennsylvania Packet* of June 3, 1777, was not exaggerating when he warned: "To all FORESTALLERS and RAISERS of the price of GOODS and PROVISIONS. Take notice that a storm is brewing again[st] you. Warning the first."[44]

A storm was indeed brewing among the craftsmen who composed the ranks of the militia. When the city militia again mobilized—this

47

time to meet General Howe's advancing army in June 1777—they marched out of Philadelphia, not to the cadence of Thomas Mifflin's patriotic oratory as they had seven months before, but to the rhythm of a new militia law that equalized participation across class lines. The Philadelphia Associators had made the passage and enforcement of such a law a condition of their service, not only because it was equitable and just, but because without it shirkers would again be free to take bread from their family's mouths. "Leaving such numbers of Disaffected in their Rear," the militiamen contended, would "render [their] Situation worse than making us prisoners of War."[45]

The storm gathered force between September 1777 and June 1778 as the British army occupied the city and rekindled hope among Philadelphia Tories and their sympathizers for a return to prewar dominance.[46] The occupation provided those who had been displaced so rudely from public affairs in the prewar years with an opportunity to regain both position and prestige under the protective cover of British arms. At the same time, this regrouping of conservative forces was aided by the sudden absence of Whigs and patriots of all ranks who fled the city to avoid capture and imprisonment.

Throughout the winter of 1777–78, as the families of exiled laboring-class patriots scratched out a meager existence in the city, elite culture flourished in Philadelphia as socially prominent loyalists and not a few moderates celebrated the restoration of "high society" in the Quaker City. In late September theaters reopened to entertain British officers and their colonial hosts. Formal balls and soirées quickly multiplied as the city's best families vied for the coveted distinction of hosting the most lavish entertainment. After a year of recrimination and vilification, the Anglophilic elite of Philadelphia breathed a collective sigh of relief and threw themselves into the task of reconstructing their social dominance with renewed energy and determination.

Yet, despite the British occupation and the resurgence of the city's elite, those craftsmen and radicals who fled from Philadelphia did not lose touch with events in the city. An underground trade in food and information was carried on between occupied Philadelphia and its hinterland from the beginning and continued throughout the occupation months. The messages that reached the exiles told of

unprecedented extravagance flaunted in the face of the difficult circumstances endured by their families and by other workingmen who
remained in the city. As unusually severe winter weather set in at the
end of October, reports filtered out of Philadelphia telling of furniture and houses being broken up for firewood and of families of
formerly respectable craftsmen being forced to beg for charity from
the city's British occupiers.

Reports of severe and degrading conditions continued throughout
the winter, and while conditions improved by early spring, laboring-
class tempers did not abate, but were inflamed further by accumulating reports of visible loyalism, suspected Quaker collaboration, and
aristocratic pretentiousness among the city's ruling class. Helpless in
their exile, the storm of discontent and frustrated action that had
been building in laboring-class minds gathered force when word
reached them of the "Meschianza" staged to commemorate the departure of General William Howe, the British commander and oc-
cupier of Philadelphia.

The Meschianza was the idea of Howe's senior officers who sought
to demonstrate their continued loyalty to a man the British were
recalling for incompetence.[47] While the Meschianza may well have
plucked up Howe's spirits, the scale and opulence of the diversion,
coupled with the enormous squandering of wealth that it represented, struck a deep chord of resentment among Philadelphia
workingmen. Patterned after the lavish masques that Ben Jonson had
performed at the court of James I, the Meschianza began with a
regatta that whisked guests along the Delaware to the Wharton estate
in Southwark. Once safely on land, Howe and his guests were first
treated to a mock jousting match performed by British officers cos-
tumed as medieval knights. This gaudy affair was followed by endless
rounds of dancing and then by a midnight supper served in a "mag-
nificent salon of two hundred and ten feet by forty," the whole affair
being lit by "three hundred wax tapers disposed along the supper
tables." The tables themselves held a feast of "four hundred and
thirty covers [and] twelve hundred dishes" for the hungry crowd, and
was followed by another round of dancing which lasted until four in
the morning.[48]

Designed as a private affair for army officers and the city's traditional elite, the Meschianza quickly became an emblem of cultural contention in Philadelphia. To the families of wealth and social standing who had been held up as objects of public derision since the summer of 1776, the Meschianza was a vindication of their way of life: the palpable extravagance and aristocratic romanticism of its medieval theme captured with unqualified precision their feelings of social superiority that had been suppressed by the rise of popular radicalism. Here, protected from popular pressure and plebeian assertion, Philadelphians of the first rank could imagine themselves in an aristocratic world where social position carried the privilege of command and deference guaranteed automatic obedience from the common classes around them.

The shipwrights, carpenters, cordwainers, and tailors who suffered in exile saw the extravagance of the Meschianza in a different light. For them, the Meschianza revealed the deep-seated loyalism as well as the corruption and moral degeneracy of the city's first families. Their open complicity in the rituals of the festival mirrored their equally open complicity with the British occupation and only reinforced, in the minds of working people, an image of the city's merchant elite as convinced aristocrats and contemptible Tories. More important, the flaunting of wealth at the festival—the wax, rather than tallow, candles; the gilded backdrops; the gluttonous amounts of food—served to sharpen the lines of class in the eyes of patrician and plebeian alike. The Meschianza, and the revival of patrician culture of which it was a part, was for the men and women of the radical-popular movement a confirmation of their worst fears about the designs of the city's elite, and it was a prescient warning about the poverty and social subordination that a British-patrician victory would bring.

The gathering storm finally broke in July 1778. The British had occupied the city for nine months and had evacuated only in mid-June, leaving Tories and moderates richer and more arrogant for the experience. The British had also left more than £180,000 in damage to the city, the bulk of it in laboring-class neighborhoods where craftsmen's frame houses had been torn down and used as firewood.[49] In the face of British destruction and Tory arrogance, a

Patriotic Society formed in late July to ferret out British sympathizers and bring them to justice. The composition of the Society was a cross-section of radicalism at mid-Revolution.[50] While outwardly similar to the city's pre-Revolutionary committees, the experience of revolution had reduced the range of wealth and occupation represented. Gone were many men of middling wealth and professional standing, and in their place sat laboring-class members of the Committee of Privates and a handful of mariners and laborers. The members of the Patriotic Society were also poorer than their pre-Revolutionary counterparts, many paying only a head tax and some having their taxes abated entirely.[51]

The Patriotic Society left no records of its affairs, but it questioned suspected Tories and collaborators and turned them over to the Supreme Executive Council for trial. In keeping with its largely laboring-class membership, the society also sought to locate those who had had a role in the destruction of workingmen's homes and to bring them to public account. But, while this was important, the true significance of the Patriotic Society lay not in the nature of its formal activities but in its bringing together of radical shipwrights, carpenters, common laborers, and mariners to pass judgment upon those who had benefited most from the occupation, and from the war itself. The Patriotic Society was a forum for militia and laboring-class discontent, and in its activities and discussions it focused that discontent onto self-interested merchants, some of whom had been their allies only two years before. The British occupation left in its wake an awareness among members of the city's laboring classes that the radical-popular alliance of the middle 1770s had natural limitations, and that, faced with powerful members of the community bent on profiting at their expense, they were increasingly alone and had only their own resources to rely on if they were ever to secure the well-being of the community and ensure their own survival within it.

We can now see the controversy over price regulation in 1778–79 and the Fort Wilson "riot" that ensued in October 1779 as the culmination of laboring-class discontent that had its roots in the years following the Seven Years' War. In the sixteen years which followed the peace treaty of 1763, artisans had mastered the workings of committees and had felt intimately the effects of war. The storm that broke in 1778 was driven by these experiences and by the selfish actions of those "instigated by the lust of avarice" who set their own private interests above those of the community as a whole.[52]

On December 10, 1778, the following item appeared in the *Pennsylvania Packet* under the signature "Mobility": "This country has been reduced to the brink of ruin by the infamous practices of Monopolizers and Forestallers [who] have lately monopolized the Staff of Life. Hence, the universal cry of the scarcity and high price of Flour." The complaint was already a common one by the end of 1778, but, unlike previous writers, "Mobility" went on to offer a solution. "It has been found in Britain and France, that the People have always done themselves justice when the scarcity of bread has arisen from the avarice of forestallers"; and he went on to remind Philadelphia's forestallers of the dynamics of the European food "riot." Several lines later, he closed with a warning that echoed, almost verbatim, the language of London tradesmen during the English civil war: "Hear this and tremble, ye enemies to the freedom and happiness of your country . . . we cannot live without bread—Hunger will break through stone walls and the resentment excited by it may end in your destruction."[53]

Since the summer of 1776 the craftsmen who supported the Revolutionary movement had received bitter fruit for their efforts. They had seen Quakers, the wealthy, and the merely better-off remain comfortably in the city while they had suffered "a Series of Hardships unusual to Citizens in private life."[54] They had watched as prices rose at ever-increasing rates, and had seen merchants and farmers hold back critical food supplies to await better prices. They had suffered the desertions of many of their middle-class supporters in the face of their democratic constitution, and had seen Tories and sympathizers treated with great leniency while they had been "calumniated and despised" by Philadelphia's "better sort."[55] And in the aftermath of the British occupation, they had found the conservative, aristocratic, anti-constitutional forces greatly strengthened and increasingly vocal and confident. In little less than three years, they had watched their hopes for a small-producer republic challenged and undercut. Their whole project lay threatened; reaction had begun.

The frustrations and betrayals, the violations of patriotic virtue, the calumniation of honest labor, the difficulty of providing for one's family: these were the preoccupations of mechanics and workingmen as Philadelphia entered its third year of revolution. The optimism and enthusiasm of 1776 had not entirely abated, but there was now an edge of defensiveness to the laboring-class voice. Depreciated currency, inflated prices, militia service, and the wartime dislocation of commerce had disrupted mechanics' trades, scattered their customers, and made a shambles of their everyday lives. In 1779 they met together, discussed the conditions into which their lives and community had fallen, drafted a reply, and nailed it onto the doors of their "Innate and Worse of Enemies": the city's forestalling and profiteering merchants.[56]

The debates of 1779 hinged on the issue of bread. Before the war, Philadelphia had been the center of a thriving colonial grain trade; by February 1779, there was not enough flour in some places to make bread.[57] It was against this background of punishing prices and the scarcity of basic necessities that a Constitutional Society formed in mid-January to mount a defense of the state's democratic constitution against conservative forces whose goal was to overturn the popular charter.[58]

The Revolutionary constitution had never been popular among city conservatives. To them, the constitution not only violated their republican principles and ideas of balanced government, but it gave power over to men whom they viewed as mere "daily drudge[s] in agricultural or mechanical labour."[59] According to Alexander Graydon, a rising young attorney and member of the militia's elite "silk stocking brigade," the framers of the constitution were a "duumvirate," a group of ignorant and wrong-headed men who had created a government that could only end in "irremediable confusion." "Nothing is republican with them," he sneered, "but as it is crawling, and mean, and candied over with a fulsome and hypocritical love for the people."[60] Even Benjamin Rush, a radical leader who took up the conservative banner shortly after passage of the state constitution, portrayed the new frame and its government as "our state dung cart with all its dirty contents."[61]

This was the frame of mind in which conservative anti-constitutionalists formed the Republican Society in late 1778 to advance the cause of a new state constitution. Organized by a small group of wealthy merchants and their sons, its leaders were men such as Thomas Willing, Robert Morris, and James Wilson, moderate to conservative Whigs who had been equivocal on the issue of separation in 1776.[62] Their pro-charter adversaries in the Constitutional Society were the popular radicals of 1776, the second tier of Revolutionary leaders who connected the city's working people to the leaders of Philadelphia's independence movement.[63] Led by William Adcock and William Thorne of the Committee of Privates, Benjamin Harbeson, a coppersmith (who, along with Thomas Paine, James Cannon, and Christopher Marshall, was a member of the city's radical "steering committee" in 1776), and George Schlosser, a German shopkeeper and tanner who had served on radical committees since the non-importation movement of 1765, the Constitutional Society carried the Revolutionary mobilization of Philadelphia's laboring classes into the 1780s.[64] There was thus a clear and direct line which ran from the political positions of the Revolutionary crisis through the competing societies of 1779. Wealth, anti-constitutionalism, and the ideology of free trade were ranged on the one side; competency, popular democracy, and the interest of the community were arrayed on the other.

By May conditions had deteriorated even further as merchants refused paper currency as payment for flour, knowing that specie-laden purchasing agents for the French fleet were due in the city.[65] In answer to ever-rising prices and the insecurity of provision among the poorest mechanics and laborers, the Constitutional Society called a town meeting for May 25 to discuss a means of lowering prices in the city.

Acting independently, the militia prepared the way for the mass meeting. On the morning before the meeting, Philadelphians awoke to find that a short broadside had been canvassed throughout the city the night before. Speaking in the direct idiom of eighteenth-century working people, the broadside reiterated a well-worn theme: "In the midst of money we are in poverty, and exposed to want in a land of plenty . . . down with your prices, or down with your selves."[66] We do not know the author of this broadside, but its closing lines point to the militia: "We have turned out against the enemy and we will not be eaten up by monopolizers and forestallers."[67] Certainly small companies of militia patrolled the city streets, driving off those who attempted to tear down the broadsides, and one company led a defacer "about on a horse bareheaded" before committing him to the city jail. The record is far from clear, but later in the day groups of militia marched through the town, collecting Tories, Quakers, and the patriarchs of Philadelphia's wealthiest families, and placed them, too, in jail. All this took place on a day when it was reported that many families were without bread.[68]

The town meeting of May 25 was the culmination of two days of militia activity. Addressed by General Daniel Roberdeau, Timothy Matlack, Dr. James Hutchinson, and a host of other city radicals, the militiamen voiced their agreement with Roberdeau, who proclaimed that "the community, in their own defense, have a natural right to counteract such combinations [of engrossers] and to set limits to evils which effect themselves."[69] To accomplish this, the town meeting elected a Committee of Trade and gave it a mandate to regulate prices beginning with an immediate fixing at the levels of May 1.

The history of popular price regulation in Philadelphia is a brief but important one. The popular-radical Committee of Trade established schedules of prices, visited and cajoled merchants, and tried to stem the flow of goods out of the city. They also acted directly to alleviate the worst problems by distributing flour and bread among the city's poor.[70] But while some merchants and retailers complied, the great bulk of the merchant community would not forgo the enormous profits that could be made by engrossing and forestalling, even if it meant near-starvation for the poorer members of the community.

In the face of this intransigence, the middle-class leaders of the Committee of Trade took an equivocal position markedly different from that of the laboring-class militia.[71] On June 5 they directed letters to city merchants and flour dealers designed to both needle their consciences and appeal to their pocketbooks. The letters ended with a warning that without bread, "discontents, far beyond our power to remedy, may probably take place," and reminded fore-stallers that "want of flour has in all countries produced that most fatal resentments."[72] The reference to the militia actions of late May is scarcely veiled, and it is clear that the radicals were themselves alarmed at the prospect of independent laboring-class action.

The committee's warnings failed to chill the blood of the city's merchants and flour dealers, however, and their letters went un-answered. The committee's response tells us a great deal about the limitations of Revolutionary radicalism. Retreating from the next logical step—calling the militia into action—the committee meekly urged the people at large to "exercise your industry [in] discovering concealed hoards" and then offered them weak advice in dealing with the city's unhumbled merchants. With a note of resignation, the committee warned Philadelphians that "you will most probably have to do with some men whose subtlety is equal to their delin-quency and who while they commit the offense, will artfully evade the punishment properly due thereto." The only advice they could offer was to "manage the best you can, always remembering the attributes of Justice and Mercy and the laws and constitution of our Country."[73] It might reasonably be asked whether this advice was addressed to the people of the city or to the merchants and dealers in provisions. It appears at once to be a call to action (but without middle-class aid) and, at the same time, a disclaimer of responsibility for whatever might ensue.

Whichever it was, on the night of June 26 the bellman cried his rounds "desireing the people to arm themselves with guns or Clubs, and make a s[e]arch for such as had sent any Flour, Gun Powder, &c. out of town."[74] Apparently, no artisans, mariners, or laborers responded; they might well have seen the call as a specious attempt to incite premature action. One thing was certain, however: such in-structions would never have come from the Committee of Trade. Facing defiance from merchants and displeasure from the laboring classes, the now-isolated committee members moved toward a solution closer to their own hearts.

It is not completely clear whether the Committee of Trade had developed the idea of price controls on their own initiative or whether they were following the lead of the militia and the Committee of Privates. What is clear, however, was that on July 26, the Committee of Trade reversed direction and offered a "citizen's plan" to stabilize the currency as a means to end price inflation. [75] This plan, while not a complete capitulation to the merchants' hard-money cure for inflation, was nonetheless an admission that price regulation had failed. [76] By September the Committee of Trade succumbed to continued forestalling, inflated prices, free-trade petitions from city merchants, and the refusal of farmers to send their food to regulated markets in Philadelphia. On the 24th the committee suspended its activities and dissolved itself as a body.

The price regulation movement failed in large part because of its inability to enforce its decisions; small merchants and artisans might be convinced or forced to follow the committee's price lists, but larger merchants, manufacturers, and farmers could not be intimidated by small groups of anxious, hungry, and angry men. Robert Morris, the greatest of the city's forestallers, found little to disturb his steady accumulation of wealth when a popular committee led by Paine, Matlack, David Rittenhouse, and Charles Willson Peale arrived to investigate his financial affairs; in fact, he felt secure enough to run a free-trade slate of candidates in the August 1779 committee elections. Ultimately, successful price regulation required popular enforcement—the cooperation of the militia and the laboring classes in enforcing price lists—yet such enforcement would also have been a direct repudiation of absolute property rights. If the Committee of Trade had called upon the militia and working people to enforce price controls in an effective way, it would have threatened social revolution and civil war.

The importance of the Committee of Trade and the price control movement lies not in its ultimate success or failure, however, but in the lines of cleavage that its failure revealed. The unwillingness of the Committee of Trade to challenge the sanctity of absolute property rights signaled the unraveling of Franklin's "strands of flax" that had bound together Philadelphia's middle-class radicals and working people for more than a generation.[77]

Evidence of this division had appeared earlier in the year with the founding of the Constitutional Society. Noticeably absent from leadership positions were those middle-class radicals who had led the struggle for independence in the city only three years before. Of the middle-class radicals of 1776 only Timothy Matlack, always the closest of the radicals to the city's laboring classes, appeared as a leader of the Constitutional Society.[78] Of the remaining Revolutionary leaders, Paine was, by early 1779, involved in provincial affairs as Secretary of the Commonwealth; Christopher Marshall and Benjamin Rush had disowned the radical movement; and Young and Cannon, the latter the author of the radical 1776 state constitution, had moved from the city.[79]

In the absence of these revolutionary leaders, Philadelphia craftsmen turned to men from their own ranks for guidance. These new leaders were mostly shopkeepers and middling artisans, some of whom, such as George Schlosser and Benjamin Harbeson, had been involved in radical activities for more than a decade. More important, most were Philadelphia militia captains, having been elected to their leadership positions by city artisans and working men.[80] With middle-class hesitation removed, working people were ready for action.

Two days after the dissolution of the Committee of Trade, the militia met on the common and reconstituted the Committee of Privates—this time without the support of middle-class radicals. Four of these radicals—Charles Willson Peale, John Bull, Alexander Boyd, and James Hutchinson—were summoned by the committee in a last-ditch effort to maintain cross-class support, but all four refused to join or to lead them. Their refusal was understandable given the purpose of the meeting; the reconstituted committee sought "to support the constitution, the laws, and the Committee of Trade" in their own way—through self-mobilization and popular action.[81]

On October 4 the militia met again on the common and began
their work. They seized John Drinker at the Quaker Yearly Meeting
and then collected Thomas Story, Buckridge Sims, and Mathew
Johns—all symbols of wealth and "Toryism"—and paraded them
through the streets "with the Drum after 'em beating the Rogues
March."[82] If the Committee of Trade could not control prices
because of "a few overbearing Merchants, a swarm of Monopolizers and Speculators, [and] an infernal gang of Tories," the militiamen declared, then the community would be better off when rid
of them. This the militia meant to do; the rogue's parade was merely
a prelude designed to show the working people of Philadelphia
the faces of their oppressors, to hold them up to public ridicule,
and to show their real weakness in the face of determined popular
might.

It was as this parade marched up Walnut Street that the incident
known as the "Fort Wilson Riot" took place.[83] James Wilson, a free-
trade merchant high on the list of popular oppressors, knew of the
militia's plans to banish Philadelphia "Tories" to British-held New
York and barricaded himself and a handful of other Philadelphia
"gentlemen" in his house to await the arrival of the militia. The
ensuing battle was a short one—one contemporary report has it
lasting, at most, ten minutes—but it was a striking portrait of the
balance of forces in Philadelphia. Wilson's house represented
wealth, profiteering, the avoidance of militia service, and opposition
to the 1776 constitution. Fort Wilson was, in short, the seat of
aristocracy and private interest. Arrayed against them, the militia
represented craftsmen, journeymen, and laborers—the working people of Philadelphia—who had experienced deprivation, loss of custom, and indignity at the hands of men like Wilson. The militia was,
on October 4, the army of artisan ideals and democratic community.
Positioned uneasily between the two combatants were the radicals—
both those who, like Peale, refused to support the militiamen and
those who, like Matlack, led the cavalry charge against them. As a
group, the radicals represented small merchants, lawyers, shopkeepers, and master artisans who disliked the excesses of Wilson as

much as they did the militia's challenge to absolute property rights. Ultimately, the radicals represented the highest bourgeois values of order, property, and social propriety.

Fort Wilson proved to be one of the hinges upon which so much of history swings. The militiamen, who were imprisoned for their role at Fort Wilson, were freed the following day, released on the bonds of their elected officers. The governing spirit in the city was one of reconciliation, and the Constitutionalists were especially eager to bring the laboring-class participants back into the party fold in order to prevent further "Tumults and outrages."[84] Wealthier Philadelphians were less sanguine, but their desires for retribution were soon contained by the prospect of conversion to hard currency. Joseph Reed, president of the state's Supreme Executive Council, was angry, but only because the militiamen had been released on local, not commonwealth, authority. By mid-November Reed asked the assembly for an Act of Oblivion, and in March 1780 the Supreme Executive Council pardoned all participants. The incident was closed—but not forgotten. The lessons of Fort Wilson were collected, transcribed, and filed in the library of plebeian culture, where they would be called upon often during the following decades.

Following the pardon of the militiamen, the gauntlet of suppression was quickly removed and the hand of reconciliation proffered by the Commonwealth government; but the resulting handshake was limp, its significance divided. To moderates and middle-class radical leaders, Fort Wilson was both a vindication of their methods and a lesson that working people were capable of thinking and acting for themselves. To city artisans it was an acknowledgment that their own organization was still weak and that moral righteousness was a small shield against the force of cavalry and cannon. To engrossing merchants and their allies, the handshake was a license to plunder, a warrant which they fulfilled with conspicuous energy and self-congratulation in the ensuing decade.

The years immediately following Fort Wilson were dark years for the working people of Philadelphia. Inflation continued unabated while congressional currency reforms served mainly to accelerate the accumulation of wealth in the hands of men like Robert Morris and

James Wilson. Forestallers and engrossers continued to withhold food and goods from city markets, and many of the laboring poor could obtain only as much food as was "absolutely necessary for family support."[85] By 1783 the cost of a typical laboring-class family's subsistence was twice what it had been eight years before.[86] Six months after Fort Wilson, Philadelphia merchant George Nelson reported that the "people seem much dissatisfied but know not how to help themselves."[87]

Yet there were positive signs as well. The radical-popular movement had indeed unraveled on October 4, but at that same moment a new charter was being written. Its preamble was drafted by a plebeian "mob" that paraded an effigy of Benedict Arnold through the streets of Philadelphia, two days *before* a march planned by city radicals;[88] by the sailors who glued paper money to their hats and to a dog who led them on a march protesting the depreciation of their pay;[89] by the militia artillery who refused to turn out against mutinous mariners who had received no pay at all; and, most presciently, by journeyman printers who demanded a rise in their "real" wages.[90] After Fort Wilson the "people" did not so much "[revert] to 'pre-political' crowd action," as one historian has suggested, but continued to enact a tradition.[91] Working people had been politicized since Keith's Leather Apron Club of the 1720s. What had happened after 1765 was that the upper ranks of the laboring classes—the independent artisans and petty entrepreneurs—had allied with the radical middle class against Tories and overbearing wealth, while those from the middling and lower ranks—practitioners in the pedestrian trades, journeymen, laborers, and mariners—had accepted (and at times sought) the leadership of middle-class and artisan radicals.

This situation changed in 1779; at Fort Wilson a line was drawn through the working community across which no sentiment or activity unauthorized by the Constitutionalists and radicals was allowed to pass. The loud and vulgar voice of the journeyman and sailor was disallowed; the rough, coarse behavior of the crowd was judged improper; the world of the laboring poor was cut away from view. But, if the poorer part of the working community was cast adrift and abandoned by the radicals, they did not come contritely, begging forgiveness. In an admittedly fitful way they worked together to articulate their own desires, to discover their interests, to argue for their particular social vision, and, in the process, to form the foundation of the city's working class.

IV

It is impossible to look back over the war years and not see a kind of rudimentary class struggle taking place in Philadelphia. But at the same time we must be careful not to read nineteenth-century experiences backward into eighteenth-century realities. The contentions between merchants, artisans, and ordinary workingmen took place in an eighteenth-century context that looked as much backward, into the village community and its marketplace, as it gestured forward toward the factory and mechanized shop.

The pressure of failed harvests and rising population had forced English monarchs to regulate local markets in food and other essential goods as early as the late Middle Ages. As historian Edward Thompson has noted, these markets were an integral part of every village community and its people held to a moral economy that gave them prior and pre-emptive purchasing rights over the produce sold there by local farmers and cottagers.[92] Much popular grievance in the eighteenth century was directed against outside traders who circumvented the traditional rights of the village marketplace by moving food directly from local farms to distant cities and foreign ports. Moral economy was, in this sense, the proclamation of a natural right to eat.

Was such a concept operative in the debate over price control in Philadelphia? Certainly, as "Mobility"'s resurrection of the food riot in 1778 illustrates, the notion of a moral economy of provision was part of the English culture shared by all Philadelphians. But while the setting of a "just price" was recommended as a remedy *in extremis* to the rising price of food, Philadelphia, unlike Boston, ordinarily experienced no episodes of popular price fixing. Town meetings and popular demonstrations in the Quaker City were concerned less with the price of bread than with inequality: inequality in militia service, inequality in wealth, inequality in personal and family sacrifice. Popular protest in Philadelphia began at that point where the English experience ended: at the point of politicization. English working people were outside of the pale of the constitution, Philadelphia working men were not. Deference and paternalism were disjointed in the revolutionary process, and without them, a moral economy of provision could not function.[93]

There was, however, another sense in which a moral economy was operative among artisans, journeymen, and laborers in Philadelphia. This form of moral economy manifested itself in the laboring-class refusal to separate moral concerns from the operation of markets and the transactions of business. It was a democratic conception of economic affairs that derived from the small producer tradition and held the well-being of the community in higher regard than the "natural" laws of supply and demand. We find evidence of this moral economy in the growing refusal of deference, in the persistent distrust of wealth, and in the positive assertion of workingmen's rights. And we find it too in the utter moral contempt with which working people countenanced forestallers, Tories, and those who made their fortunes at the expense of others. The conflict of power and opinion that took place over the issue of price regulation separated those who held this notion of moral economy from those who reserved the exercise of morality for more private, non-economic affairs.

These competing economic ideas took on a public mien in September 1779, when merchants and workingmen debated the virtues of price control in the pages of the *Pennsylvania Packet*.[94] Hounded by the Committee of Trade and faced with mounting opposition from artisans and other workingmen, prominent city merchants had met in late August to draft a public response to the popular price-control movement. In their arguments against the ethics of price control, the merchants embraced the emerging economic liberalism of the trans-Atlantic trading community. Taking their lead from Adam Smith's *Wealth of Nations*, copies of which had arrived in the city shortly after its publication in 1776, the merchants began by elevating Smith's invisible hand to the status of natural law. The unfettered operations of the marketplace, they declared, "have their foundation in the laws of nature, and no artifice or force of man can prevent, elude, or avoid their effects."[95] The law of supply and demand was, in their words, "uncontrollable" and any attempt to impose price controls was therefore doomed to failure. Sensing that the harsh judgment of the marketplace might appear overly callous and provocative to city artisans, the merchants finally turned to simple utility in support of their case. "Freedom of Trade," they explained, "or unrestrained liberty of the subject to hold or dispose of his property as he pleases is absolutely necessary to the prosperity of every community, and the happiness of all who compose it."[96] Seen from the merchants' viewpoint, price controls were not only morally reprobate and impractical, but they worked against the economic interests of the community as well.

The popular radicals thought differently. As early as 1776 they had supported a constitutional provision limiting the accumulation of wealth and property. By the summer of 1779 the experience of engrossed goods, high prices, deflated currency, and encroaching poverty led them to articulate not only the grievances of the city's laboring classes but their prescription for a moral community as well.

The social philosophy articulated by the radicals was in direct contrast to the liberal notions of the free-trade merchants. Where the merchants had envisioned "community" as a collection of independent individuals, the radicals conceived of it as a network of social relationships which alone allowed men to pursue their trades and live out peaceful and productive lives. To the radicals of the Committee of Thirteen—men such as Thomas Paine, Timothy Matlack, and Charles Willson Peale—"right in the community" was fundamental:[97]

> The social compact or state of civil society, by which men are united and incorporated, requires, that every right and power claimed or exercised by any man or set of men, should be in subordination to the common good, and that whatever is incompatible therewith, must by some rule or regulation, be brought in subjection thereto.[98]

Harkening back to small-producer ideas of the primacy of community and the rights and benefits derived from it, the Committee of Thirteen took up the fundamental issue that separated producer and merchant. According to the merchants' manifesto, price control was inherently unjust because "it invades the laws of property."[99] But what was property? The radicals set about providing an answer by making a critical distinction between the concepts of "property" and "service."

The Committee of Thirteen began by asking Philadelphians to consider whether any merchant had the right to set prices at will or to monopolize an essential commodity while waiting for the price to rise. Arguing that he had not, the committee reasoned that, as the merchant could not possess any articles

> without first deriving assistance from the collected efforts of the community, it follows as an act of duty, that having first received advantage from their service he owes them his in return, at a price proportioned to what he gave; and therefore, whenever the avarice of individuals occasions them to transgress this line, the principles of public justice and common good require it to be limited.[100]

Service, or the "collected efforts of the community," thus placed a prior claim on the property of the merchant. The working people of the city had given their service, in labor, to the merchant, and they should now expect his service, in the form of fair merchanting, in return, irrespective of the legal "ownership" of the goods. In the notion of service, then, we find a reciprocal obligation on the part of every community member to exercise his own calling toward the collective good. This notion was further clarified as the Committee reviewed the history of the preceding twelve months:

> [After the British withdrawal in September 1778] there was not a vessel, scarcely a boat, to be seen in the river. It was therefore impossible that those who then professed themselves merchants . . . could exercise their professions without the accumulated assistance of the several trades and manufacturers concerned in the art of building and fitting out vessels for sea. Ship carpenters, joiners, blacksmiths, gunsmiths, blockmakers, tanners, curriers, painters, and labourers of numerous

kinds, contributed their several portions of service to this purpose. When the vessel was on float and capable of sailing, another set of men were employed to victual her, and a third to man her; and without the previous assistance of all these, the merchant would have been only an unserviceable name applied to an occupation extinct and useless.[101]

From this argument emerges a notion of community as a collective enterprise in which men might be divided by the form of their labor, but in which they were not the less equal for it. The differences between the ship carpenters who built the ship, the victualers who provisioned it, and the mariners who sailed the vessel were of no more consequence than the differences in occupation of the merchant and the workingman. Men were to be judged by their contribution to the community, each in their own capacity, and not by the nature of their occupation itself. As the argument makes clear, the merchant's function, or serviceability, within the community was made possible by the efforts of the workingmen who built, victualed, and manned the merchant's vessel. His position as a merchant, as an owner of goods, or as an employer of labor gave him no special privilege either with regard to the vessel or within the community as a whole.

Anticipating an argument from merchants and employers who viewed labor as a simple commodity, the Committee of Thirteen asserted that the payment of wages conferred no special authority on the merchant or employer, nor did it empower them with exclusive control over vessels or the products in trade:

We conceive that these men [the builders, victuallers, and mariners] and all others concerned had something more in view than their meer wages when employed in the constructing and fitting out vessels for mercantile purposes, and that they naturally considered themselves as furnishing such vessels, not so much for the particular emolument of the merchant who employed them, as for the more beneficial purposes of supplying themselves and their fellow citizens with foreign necessities; and therefore we hold, that though by the acceptance of wages they have not, and cannot have any claim in the *property* of the vessel, after it is built and paid for, we nevertheless hold, that they and the state in general have a right in the *service* of the vessel, because it constitutes a considerable part of the advantage they hoped to derive from their labours.[102]

Summarizing their counterattack on the merchant claim of absolute property rights in goods and labor, the Committee of Thirteen wrote that "the *property* of the vessel is the immediate right of the owner, and the *service* of it the rights of the community collectively with the owner."[103] It was a seemingly contradictory idea. On the one hand it attacked the whole concept of inviolable property rights that was common currency among the wealthy and educated classes, while, on the other, it conceded the merchant's right to "own" the vessel itself. But the idea was, in fact, less contradictory than it at first appears. The Committee of Thirteen was attempting to articulate the fundamental principles of a form of society the rudiments of which had existed among Philadelphia artisans since 1681. In their notion of the primacy of community and the service each individual owed to it, the Committee was in 1779 describing the philosophical underpinnings of a small producers' society that had illuminated the left wing of the Leveller movement 130 years before.[104] For nearly a century, the working people of Philadelphia had been living in just such a society, and it was because they had felt that society threatened and attacked by an expanding commercial capitalism that they formed committees of defense and began to work out the meaning of their own ideas of community. Over the course of the next half-century, artisans would refine their notions and organize themselves in unique ways in order to keep Philadelphia a community of small producers. And in the nineteenth century, the notion of service would be recognized for what it was—a labor theory of value—and would form the cornerstone of the working-class movement. Once this theoretical problem was solved, the way was open for a more powerful argument to be made against the dominance of private property rights; only then could the ideal of a community of productive workers be built on a more substantial moral foundation.

There was something fundamental being worked out in 1779. It was the eighteenth-century moral economy, with its history running back to the village community, being wielded and tested against an emerging capitalist ideology. And in that battle, what was a coarse, blunt, and unwieldy instrument was transformed—an illusion chipped off here, a clearer notion fashioned there—into a sharper and more efficient instrument of struggle.

CHAPTER 3

Reaction and Restoration:
Artisan Politics in Crisis,
1780–1789

The defeat of the militia at Fort Wilson was experienced by the whole of Philadelphia's artisan community as well as by the militiamen directly involved. The troops who rode to Wilson's defense were pelted with stones and refuse, and unknown persons were rumored to be plotting the release of the militiamen from the city jail.[1] It takes little imagination to see the quick release and pardon of the prisoners as an attempt by state and local officials to forestall further demonstrations and to defuse community unrest. Timothy Matlack, the newly appointed Secretary of the Commonwealth, confirmed the depth of community support for the militiamen in a letter to Chief Justice Thomas McKean. "What will be the event of tomorrow is very uncertain," he wrote, "as the people are heated to an extreme degree."[2] The letter was dated October 6, 1779, two days after Fort Wilson and a full day after the release of the militiamen on the bonds of their officers.

Both the fact of independent popular action and Matlack's uncertainty about laboring-class sentiment bear witness to the distance that separated the city's radical leaders and their former laboring-class allies. From October 1779 forward we can trace two histories of popular radicalism in Philadelphia. That of the radical leaders is well-known: Matlack, Rittenhouse, Hutchinson, Peale, and McKean became leading Constitutionalists, and turned their energies to government. Their history is the story of party contention, parliamentary maneuvering, and the channeling of dissent. In the end, some, such as Peale, found the rough-and-tumble of politics distasteful and retreated into the comforts of private life; but others, such as McKean, remained party leaders throughout the 1780s.

There is a second history, less well-known, that deserves greater attention. It is a history of nameless cordwainers, journeyman printers, and apprentice cabinet-makers; a record of intimate discussions and personal allegiances that can be measured only in their results; an account of ordinary working people as they sought to establish an honorable place for themselves in an uncertain post-Revolutionary world. The working people of Philadelphia did not retreat into apathy after 1779; rather, they faced the political crisis of the 1780s by turning to themselves—a Mechanics' Committee formed here, a journeyman's society meeting there—to give their own truth to the meaning of independence.

I

The public issues of the 1780s were so closely imbricated with the workings of factional politics in Philadelphia that it is often difficult to distinguish honest sentiment from partisan invective. Yet a few things are clear. In the immediate postwar years, both Constitutionalists and Republicans accepted the central republican principle that government derived its authority from the will of the people and both actively sought to secure their share of the city's popular vote. Yet, at the same time, both parties also agreed with Benjamin Rush, a one-time radical turned Republican spokesman, when he wrote: "it is often said that the sovereign and all other power is seated *in* the people! This idea is unhappily expressed. It should be—all power is derived *from* the people. They possess it only on the days of their elections. After this, it is the property of their rulers."[3] Neither Republicans nor Constitutionalists were democrats, and the leaders of both parties shared a common view of themselves as the natural rulers of society by virtue of their social position, their education, and their cumulative experience in government.

Beyond this simple commitment to representative government and a broadly defined republicanism, however, the two parties disagreed about everything else. From the beginning, Republicans and Constitutionalists thought of each other only in hyperbole, and the reader of the city's partisan press soon learned to recognize the competing parties in the epithets "despot," "aristocrat," and "skunk." Yet beneath the spleen and screed was also a sense that the Republican and Constitutionalist parties represented serious and conflicting principles. If the Republicans were not exactly ready to reinstitute a monarchy and House of Lords, as their opponents occasionally claimed, they did want a more authoritarian state constitution in which the interests and influence of the wealthy would be permanently institutionalized.[4] And if the Constitutionalist propaganda sometimes overshot its mark, it nonetheless caught the truth of postwar city politics; for what the Republicans sought was nothing less than a Restoration; not of king and parliament, but of the rule of wealth and "family" that had existed before the Revolution and which already prevailed in most of the remaining states of the Confederation.

While lacking the focused program of their opponents, the Constitutionalists nonetheless represented a radically different view of government. As their party label proclaimed, the Constitutionalists identified themselves with the state Constitution of 1776, which they saw as essentially sound, if in need of periodic revision. More important, they saw it as an instrument that kept power where it properly belonged: close to the people of Pennsylvania. In a state whose inhabitants were chiefly small farmers and craftsmen, their views were in large part correct: the radical constitution was a frame of government designed for a society of small producers. The opposition between Constitutionalists and Republicans was in this respect a conflict between two notions of Pennsylvania's future. All of the issues, rhetoric, charges, and counter-charges of the era came ultimately to be measured against this standard: a choice between, on the one hand, a rushing, grasping, commercial society wedded to an ideology of the main chance and, on the other hand, a more measured producer's society where human cycles, and not business cycles, claimed the community's allegiance.

The Republican Restoration was worked out over the space of a decade and focused on four related issues: revision of the Constitution of 1776; the transfer of commercial regulation from individual states to the national government; a defense of the Bank of North America; and the restoration of the municipal corporation in Philadelphia. The Republican program was thus a conscious and coordinated effort to undo the radical work of the Revolutionary committees and to reinstate the dominance of Philadelphia merchants and their backcountry supporters in the economy and politics of Pennsylvania. It was assumed that their program would, of necessity, incorporate and discipline the artisans and workingmen of Philadelphia.

Republican hopes for a restoration rested on the drafting of a new state constitution.[5] The Constitution of 1776 was, in their eyes, too democratic; more to the point, it made no institutional provision for men of large property, either in the form of an upper legislative house or restrictive property qualifications for office-holding. Assaults against the radical constitution had begun in the conservative resurgence fostered by the British occupation of 1777–78. By 1779 the conservative Republican Society was publicly calling for a convention to draft a state constitution that would include both a single executive and an upper legislative house. Faced with unbending popular support for the Revolutionary constitution, this first Republican bid for revision met a resounding defeat in the October 1779 elections. But while they had lost the initial skirmish, the revisionist cause was not long forgotten.

The Republicans registered increasing gains throughout the remaining war years and won a slim legislative majority in the 1782 elections. But the real contest over the constitution took place the following year and focused on the election of a Council of Censors, a body with the constitutional authority to amend the frame of government or call for its redrafting.[6] The hard-fought censorial election of 1783 gave the Republicans a simple majority on the council, and with this numerical advantage in hand, the conservatives set about reviewing the constitution and all laws passed under its aegis. By early 1784, they had finished their work, and on January 17 council Republicans issued their majority report which, as expected, condemned not only the unicameral legislature but the state's plural executive, the public accountability of judges, and statutory rotation in office as well. Each of these condemnations struck at what the Republicans considered to be the constitution's overly democratic provisions, provisions which in their eyes had served to undercut authority and had led to abuses from below. With the issuance of this report, the Republicans began an attack on the radical constitution that would occupy them for the remainder of the decade.[7]

The Republican challenge did not go unanswered. Less than two weeks after the appearance of the majority report, the Constitutionalist minority on the council published a direct appeal to the people of Pennsylvania in which they reminded their constituents that the calling of a constitutional convention required agreement by two-thirds of the council, a condition that the Republicans, with only a 12 to 9 majority, did not meet. The minority appeal went well beyond legalistic quibbling, however. Printed in newspapers across the state, the minority warned that if the Republicans succeeded in abrogating the Constitution of 1776, the results would be a form of political slavery that could only be undone by "manur[ing] the Plains of a second Runnemede with the richest blood of America."[8]

The minority report carried this equation of Republican design with aristocratic government throughout their appeal, at one point admonishing Pennsylvanians to ignore the "artful addresses of these aspiring despots, who wish to establish an upper house of lords amongst you, that they may thereby more effectually teach you submission to your *betters*."[9] But beneath the hyperbole was a more serious analysis that struck at the heart of the Republicans' restoration program:

> The grand objection to our present constitution is that it retains too much power in the hands of the people who do not know how to use it so well as gentlemen of fortune, whose easy circumstances give them leisure to contrive how to spend your money to the best advantage, and that it gives no advantage to the rich over the poor.[10]

The minority address brought Republican convention hopes to a halt. By February 1784, remonstrances against the Republican report circulated throughout the state and it rapidly became clear that the people were prepared to fight for their radical constitution.[11] Faced with certain defeat, on September 15 the Republican-led council declared "that there does not appear . . . an absolute necessity to call a convention, to alter, explain or amend the Constitution."[12]

This was not the end of the conservative efforts, however. Outmaneuvered by the Constitutionalist's successful public appeal, the Republicans turned their full attention to the legislature and sought to gain their ends through the legal and parliamentary maneuverings they knew best. Faced with volatile public opinion and uncertain popular support, Republican leaders recognized that the success of the Restoration now rested on their ability to attract new voters to their cause. They did not have far to look. Between 1777 and 1779 the revolutionary state government had passed a succession of test acts that disenfranchised anyone who refused to swear or affirm their allegiance to Pennsylvania's Revolutionary government. Foremost among those affected were the state's Quakers, who because of religious scruples against oath-taking or antipathy toward the Revolutionary movement—and often both—had been excluded from Pennsylvania politics for nearly a decade. Angry and frustrated by their political exile, the Quakers became the key to an unassailable Republican majority, and in 1786 the Republican-controlled legislature moved to incorporate this vital bloc of disenfranchised voters by annulling the controversial Test Act. By the end of the decade, the votes of re-enfranchised Quakers brought the Republicans the success they had worked so hard to achieve.

The calling of a state constitutional convention in 1789 was the culmination of fourteen years of vigorous effort on the part of the Republicans. Only public reaction to their attempt to abrogate the constitution through the Council of Censors in 1784 had interrupted their record of slow, but steady, gains in the state legislature. In the final years of their campaign against the Revolutionary constitution, the Republicans held power in the legislature, but moved cautiously, introducing their revisions piecemeal in order to avoid a renewal of the public furor of 1784. With the ratification of the federal Constitution in 1788, the Republican position gained added strength from the similarity of their proposed state constitution to the frame of the new national government. Assured of victory, in September 1789 the Republican assembly finally declared, over waning Constitutionalist objections, that a convention was the true will of the people.

The new frame of government that was ratified in September 1790 erased the democratic provisions of its revolutionary predecessor by removing most aspects of government from immediate popular control.[13] Not only did the revised constitution divide legislative authority between an upper and lower house, it removed most provisions for rotation in office and shifted the balance of power away from the legislature, placing it firmly in the hands of a single executive. Unlike the limited powers of the Supreme Executive Council, Pennsylvania's governor now had the power to appoint not only the secretary of the commonwealth, the attorney general, and the state treasurer, but the entire commonwealth judiciary as well. The appointive powers of the governor even extended into the local community, where hundreds of lesser officers now owed their jobs to executive appointment. Of the innovations established in 1776, only the provision for manhood suffrage survived the constitutional transformation, albeit with an extended residency requirement that kept newcomers from the polls for an additional year. Taken together, the provisions of the Constitution of 1790 provided the institutional, legal, and political base for the restoration of elite rule in Pennsylvania.

Yet while the new constitution guaranteed to the wealthy a powerful voice in government, it could not, in itself, solve their most pressing problems. Wealth in Pennsylvania was centered in Philadelphia, and wealth in Philadelphia came predominantly from commerce. When war contracts ended in 1783, Philadelphia merchants turned their attention to the West Indies trade, which had been the lifeblood of the city's economy since the early eighteenth century. Foremost in their minds was a commercial treaty with England that would open British West Indian ports to their ships and goods. From the beginning, city merchants recognized that this could best be accomplished by the national government rather than by individual states, and in the years from 1783 until the adoption of the federal Constitution in 1788, the Republicans made the nationalization of commercial regulation the principal aim of their economic policy.

The Republican campaign began in late 1783 with two pamphlets written by Thomas Paine and William Bingham, a prominent Philadelphia merchant, that put forward the American case for renewed West Indian commerce.[14] Then, in early January 1784, the Philadelphia Merchants' Committee wrote to their counterparts in Massachusetts to suggest united action in Congress. But despite this early agitation, little was done to shift the control of international trade to Congress until 1785, when depression set in and Philadelphia's economy slowed to a snail's pace. By summer "One of the People" reported that "rents are falling, tradesmen can hardly be employed [and] when employed they are not paid; most people are running in debt."[15] Two weeks later, the editor of the *Pennsylvania Packet* claimed that a "greater number of people have taken shelter in the new gaol, in order to 'clear off their scape' by the benefit of the new insolvency acts than have been ever before known in this city."[16]

On April 6, 1785, the Merchants' Committee memorialized the assembly, arguing that the fundamental defect in the national Congress was its lack of control over trade. Along with this continued pressure on the assembly, the Republicans added their first appeal to the laboring classes.[17] The postwar depression of 1785 struck with particular severity at craftsmen and laborers in the maritime and allied trades. In response the Republicans shifted the emphasis of their campaign away from commercial treaty negotiations to a new focus on the protection of American producers. The transfer of commercial regulation to the national government was offered as a panacea for the afflictions of depression; not only would Congress have the prestige and authority to bring about a successful commercial treaty with Britain, but such a treaty would be national in scope. This national focus would, according to one Philadelphia editor, obviate the problem of "each state always counteracting another."[18] Moreover, if Congress was given control over commerce, the merchants argued, uniform tariffs could be established and the manufacturers of the entire Confederation protected from the avalanche of British goods that were in some cases selling for less than the domestic cost of production. One contemporary observer summed up the merchant argument cogently: for the want of congressional power to regulate trade, he wrote "we can never be happy."[19]

From 1785 until the fall of 1787 Philadelphians appeared to be united in their support of a strengthened national government able to deal effectively with commercial regulation. But as rumors about the nature of events taking place in the closed constitutional convention began to be confirmed in late summer 1787, and the outline of the new Constitution became increasingly clear, an opposition movement began to form. Speaking for a group of Constitutionalist legislators in early October 1787, George Bryan, a member of the state's Executive Council, alerted the people of Pennsylvania to the dangers of the Republican plan. The state legislature, he recalled, had originally appointed delegates to the Philadelphia convention with "the purposes of revising and amending the present Articles of Confederation."[20] But, he noted, no sooner had the convention's preliminary proposals been received by the Republican-controlled legislature, than they called for a second convention designed to abrogate the Articles of Confederation and fix a new frame of government in its place. Outnumbered by assembly Republicans, Bryan recalled that the Constitutionalists had attempted to delay proceedings by denying the House a quorum. But, undaunted, on the morning the question of a convention was scheduled for debate "a number of citizens of Philadelphia" descended on the lodgings of two backcountry Constitutionalists and, according to Bryan, "violently [broke] open" their rooms, tore their clothes, "and after much abuse and insult, forcibly dragged [them] through the streets of Philadelphia to the State House."[21] In this way the Republicans secured a quorum and proceeded to vote in favor of a constitutional convention.

Bryan made no comment on the incident, but one was hardly needed. The secrecy, precipitancy, and violence employed by the Republicans was argument enough and revealed in unmistakable terms the nature of their political designs. As public debates over the new constitution began in November 1787 the two political parties, now more commonly known as Federalists and Anti-Federalists, marshaled their forces and mounted the most heated public campaign that Philadelphia had witnessed since the Revolution.

The resulting exchange found one of the radical leaders of the 1770s squarely at the center of the controversy. Writing as "An Old Whig," Dr. James Hutchinson addressed himself to the small producers of Philadelphia. [22] Echoing the sentiments of the majority of Philadelphians, Hutchinson declared that he was "one of those who have long wished for a federal government which should have power to protect our trade."[23] And, like many Philadelphians, he admitted that he "was disposed to embrace [the Constitution] almost without examination."[24] But upon closer inspection, Hutchinson had found several points of concern. Most Philadelphians, he wrote, saw the Constitution as an experiment: "We want something: let us try this; if it does not answer our purpose we can alter it."[25] But, he asked, what kind of experiment was it when the proposed Constitution was so difficult to alter? By making Congress solely responsible for initiating amendments and by requiring agreement of two-thirds of the state legislatures before any amendment could be approved, Hutchinson argued, "the constitution . . . never can be altered or amended without some violent convulsion or civil war."[26]

Hutchinson cautioned the people of Pennsylvania not to "be in haste [and] consider carefully what we are doing."[27] The dangers were great, he warned, for "the new constitution vests Congress with such unlimited powers as ought never to be entrusted to any men or body of men." If the Constitution were adopted, he argued, "no other power on earth can dictate to [Congress] or control them, unless by force."[28]

Turning his appeal to those workingmen with the most to gain from the commercial clauses of the Constitution, Hutchinson declared that while it was the case that "our shipwrights are starved, our seamen driven abroad for want of employ, our timber left useless in our hands, our ironworks . . . almost reduced to nothing, and our money banished from the country," it would be tragic if, for immediate relief, workingmen supported a new government that departed from "the principles of liberty" embodied in the Articles of Confederation. "Scarcely a man of common sense can be found who does not wish for an efficient federal government," he wrote, but while "power is very easily encreased . . . it hardly ever lessens."[29]

Hutchinson's essays were directed primarily to those supporters of the Constitutionalist party who were indeed "Old Whigs." Despite his gesture toward mariners and maritime craftsmen, Hutchinson's arguments were, in truth, a measure of the distance that separated Philadelphia workingmen from the Constitutionalist leadership in the late 1780s. If the Constitutionalists, or Anti-Federalists, wanted the support of the city's laboring classes, they would need more than Whig rhetoric; they needed a political program that would relieve the widespread distress felt by Philadelphia's working people.

The first step in this direction was taken in October 1787 when "Montezuma" addressed the city's workingmen, not in the language of republicanism, but in simple class terms. The Federalists, he mockingly observed, were "the Aristocratic Party of the United States" and had long lamented "the many inconveniences to which the late confederation subjected the well-born."[30] Not the least of these "inconveniences," he explained, was that "the *better kind* of people" had been brought "down to the level of the *rabble*."[31] Now, with the federal Constitution in hand, they were poised to regain their former privileged position in society and planned to institute a "*monarchical, aristocratical democracy*" to ensure the preservation of their power. [32]

The Federalist trick, "Montezuma" informed laboring-class Philadelphians, was "to indulge in something like a democracy in the new constitution [to be] designated by the popular name of the House of Representatives." "But to guard against every possible danger from this *lower house*," he continued, the Federalists had created an upper house and a single executive to counteract "that *grand engine* of popular influence." Moreover, in order to secure their rule even further, "Montezuma" argued, the Federalists placed the upper house beyond the realm of popular election, thus "prevent[ing] the representatives from mixing with the *lower class* and imbibing their foolish sentiments."[33]

"Montezuma"'s irony and his thinly veiled comparison of the elitist federal Constitution to the more popularly oriented state constitution of 1776 set the tone for the "Address of John Humble" in October 1787.[34] Posed as a "humble address of the *low born* of the United States of America to their fellow slaves scattered throughout the world," Humble's address harkened back to the language of popular protest during the English civil war as well as that of Philadelphia's popular political awakening in the 1720s.[35] "Although our sensations in regard to the mind be not so nice as those of the *wellborn*," he declared, "yet our feeling, through the medium of the plow, the hoe, and the grubbing axe, is as accute as any nobleman's in the world."[36] What the John Humbles of Philadelphia saw was that the well-born, "in the profundity of their great political knowledge," had agreed that "nothing but a new government consisting of three different branches, namely, *king*, *lords*, and *commons*, or in the American language, *President*, *Senate*, and *representatives*, can save this our country from inevitable destruction." Though "several of our *low-born* brethren have wickedly began to doubt concerning the perfection of this evangelical constitution," "Humble" wrote, they were being pressed by Federalists to declare "without any equivocation or mental reservation whatever" their support for the new Constitution.[37] The result, he warned, would be a new "*Royal* government [supported] by the sweat and toil of our bodies"; a government which would maintain a standing army, increase taxes, demolish trial by jury, and suppress freedom of the press. In the end, "all *power*, *authority*, and *dominion* over our *persons* and *properties*" would be left "in the hands of the *well born*, who were designed by Providence to *govern*." Under the new Constitution, "Humble" concluded, small producers would be left powerless, "contented if our *tongues* be left us to lick the feet of our well born masters."[38]

In the ensuing months, as the Constitution approached ratification, Federalists followed the lead of Anti-Federalists such as "Montezuma" and "Humble" in appealing to the city's laboring classes for support. Taking the name of a rising laboring-class district, "Southwark" compared the Anti-Federalists, whom he claimed denied "the universal distress and complaints of the people," to the Tories of the Revolution.[39] Two weeks later an anonymous writer to the *Pennsylvania Gazette* noted the importance of the laboring classes to the city's Federalist cause when he commented "that the uniform of the Federalists in this city is to consist of cloth covered buttons, leather pockets, and plain shirts."[40] In the same issue of the *Gazette*, another Federalist declared himself to be "One of the People" and proclaimed that "the interests of . . . the farmer, the mechanic, [and] the merchant . . . are intimately blended together." The result of the Confederation, he went on to argue, had been "bankrupt merchants, poor mechanics, and distressed farmers," and only government under the new Constitution would be "conducive to the happiness and . . . interest" of the people.[41] Turning directly to the workingmen, "One of the People" referred to the glut of imported British manufactures in the city and asked, "whence comes it that shoes, boots, made-up clothes, hats, nails, sheet iron, hinges, and all other utensils of iron are of British manufactory?"[42] His answer was predictable: this ruinous situation was the result of the weak Confederation government. With adoption of the Constitution, he promised, past calamities would be turned aside and "our mechanics will lift up their heads and rise to opulence and wealth."[43]

This was the core of the Federalist appeal: a strong national government would implement a comprehensive tariff, revive American commerce, stem the flow of British goods into Philadelphia, and bring prosperity to all.[44] It was a compelling argument, especially for those craftsmen who faced ruinous competition from imported goods, and many laboring-class Philadelphians accepted it. Quaker City craftsmen were a prominent force at the town meeting of September 20, 1787, that issued the first public call for adoption of the Constitution.[45] And soon thereafter, one disheartened Anti-Federalist reported that "the people in this city are very anxious to have [the Constitution] adopted" and begrudgingly noted the success of Federalists who were "running into the country and neighboring towns haranguing the rabble" to join them.[46]

Federalists were confident in the fall of 1787 and expected the city's laboring classes to fall easily into step on their march toward ratification. And, for the most part, historians have followed their lead. Reading Federalist claims and glancing backward from the

Federal Procession of 1788, which found Philadelphia craftsmen marching with the emblems of their trades in celebration of ratification, historians have concluded that the city's laboring classes donned the cloak of Federalism in the late 1780s.[47]

The reality of those years is more complicated than this. It is clear that mariners and practitioners of the maritime crafts supported the Federalists. Whether they did so from pure economic interest or from the personal ties of patronage that bound them closely to Federalist merchants and manufacturers makes little difference; modern historians and contemporaries agree that these working people were firmly in the Federalist camp.[48] But what of the city's craftsmen who were not directly tied to commerce and shipbuilding? The evidence is scattered and fragmentary. Federalists were indeed victorious in the 1787 assembly elections and in the November 1787 election for delegates to the state ratifying convention.[49] They could not have won this victory without the support of a substantial part of the laboring community. But these two elections were held within six weeks of the public reading of the Constitution on September 20, before substantial opposition could be organized in the city. Thus in Philadelphia, as elsewhere, speed was crucial to Federalist success. Federalists everywhere pressed for short and controlled discussions of the Constitution and speedy ratification by the states in the hope that they could gain their end before an effective opposition movement could mobilize potential dissent.[50]

During this same six-week period, when organization was needed, Philadelphia Anti-Federalists had little to offer. The city's leading Anti-Federalist, Dr. James Hutchinson, may have authored the "Old Whig" essays, but he was also one of those who had refused to lead the militia at Fort Wilson in October 1779. He had spent the 1780s slowly trying to rebuild the radical-popular alliance that his refusal at Fort Wilson had helped to shatter. But while he had achieved some success in the mid-1780s through his party's support of a mildly protective tariff, in 1787 Hutchinson could by no means claim the allegiance of the bulk of Philadelphia's laboring-class voters.[51] What the Anti-Federalists needed was time.

When the Pennsylvania convention met in November 1787 ratification of the Constitution was a foregone conclusion. Left with little else, the Anti-Federalists focused their attention on limiting the powers of the new national government through the adoption of a bill of rights. And with this new tack, anti-Federalists began to garner laboring-class support. In April 1788 "A Mechanic" owned that he was "one of that class of citizens who suffer as much by the dullness of the times, and scarcity of cash, as most."[52] But, he declared, "neither am I carried away with the bombastic or artful declarations [of the Federalists] who tell us that [the Constitution] will not only relieve all our complaints, but cure all our sores."[53] "Such *Canterbury Tales*," he wrote, did not hide the fact that "without either a *bill or rights or representation* and with a standing army," our rights are insecure.[54] With considerable prescience "Mechanic" went on to anticipate popular complaints against Washington's "Republican Court" of the 1790s when he argued that the new Constitution "without amendments, far from relieving our distresses, would increase them tenfold by the enormous taxes which must necessarily be laid to support a superb presidential court and numerous list of officers."[55]

"Mechanic"'s essay is the first public indication of growing laboring-class Anti-Federalism in Philadelphia. As he observed, "For some time the infatuation was great, but the tide of political frenzy has reached its height and [is] ebbing fast towards the beach of policy and reason."[56] In the end, "Mechanic" had "no doubt but that a great majority of the freemen of Pennsylvania are opposed to the chains forged for them . . . and will act as a majority should when their liberties are at stake."[57]

"Mechanic" spoke for those working people who had formed, in the words of one Anti-Federalist leader, an "attachment to the Revolution & Constitution of Pennsylvania."[58] The late 1780s was a time of division among the city's laboring classes. Their Revolutionary leaders had deserted them or had left the city altogether, and the Constitutionalists, dominated by a powerful rural constituency, had established only very tenuous and unreliable links with workingmen. Laboring-class Philadelphians faced the constitutional debates of 1787–88 with a desire for change but without their own program or the unity necessary to bring it about. For their part, the Federalists offered a reasoned program and mounted a well-organized and well-funded campaign in support of it. That so many workingmen, faced with the everyday experience of growing impoverishment, saw the Constitution as salvation is less a sign that they had taken the side of the rich and powerful than an indication that they sought the best form of relief available. As a Philadelphia "Bricklayer" put it, "an honest mechanic who serves his country faithfully is as well deserving of her favor as another."[59] A vote for the Constitution was for him a vote for a tariff, and that was, in turn, a vote for a normal working life. The fact that workingmen began to follow the Anti-Federalists over the issue of a bill of rights and that they would turn with such speed and conviction against Federalist grandees in the early 1790s throws into doubt the commonly accepted notion that Philadelphia's laboring classes were staunch Federalists in the late 1780s. What is more likely is that they saw a national tariff as a simple question of survival and voted for the Constitution that they hoped would bring it into being.

With the ratification of the Constitution in July 1788 the power of commercial regulation was firmly established at the federal level and the Republican Restoration attained one of its crucial economic goals. The next step was to devise a national system of credit to secure the financial position of the wealthy.

In the late eighteenth century, international trade required ready access to credit, often in substantial sums and for extended periods of time. Lacking sound financial institutions, American credit after the war, as before it, was extended and controlled by Britain. Vexed by the expense and general difficulty of obtaining British credit, Robert Morris and the Republicans sought to free the merchants and manufacturers of Philadelphia from English financial imperialism by creating the Bank of North America.[60] Chartered by Congress in 1781, the bank was initially designed as a national depository to handle the wartime finances of Congress. Its stockholders were limited to the original subscribers (mostly the family and friends of Robert Morris) and it paid handsome dividends. In addition, the bank's monopoly

on credit gave its directors virtual control over business in Philadelphia, extending credit to those it favored and denying it to those it
did not.

Constitutionalists, especially those from rural counties where
farmers could ill afford the bank's lofty interest rates, objected to the
bank from the beginning. By the mid-1780s complaints against the
bank had become common and ranged from the favoritism shown
toward Morris's friends to its potential for subverting the Republic.
One observer provided a balanced analysis of the causes for discontent in 1784:

> The present bank has been both useful and pernicious: useful to the
> stockholders, who got a prodigious interest for their money; and useful
> to the man of trade, because, by paying a moderate discount he could
> turn his notes of hand into cash. But it has been prejudicial to the
> middling and poorer classes of people who wanted money on interest;
> for, as the bank yielded 9, 12 and even 16 per cent, none could be had
> at the legal rate, which is 6, and many were much distressed for want of
> money who could give good security, but could not afford to pay so
> great an interest. The brokers took the advantage of people's necessi-
> ties, and have let money out by the month, at 4½, 5, and even 6 per
> cent per month, which is 54, 60 and 72 cent per annum. Such folks
> can always find means to evade the law.[61]

Double and treble interest along with the blatant preference
shown to Morris's wealthy and socially prominent associates brought
farmers and craftsmen together in opposing the bank and forced
them to seek an alternative means of credit. In their search for a more
democratic credit institution, many of the bank's opponents turned
to the example of the General Loan Office which had operated
successfully in Pennsylvania from 1723 until the beginning of the
Seven Years' War.[62] The Loan Office was a government-run bank
that printed money and loaned it at 5 percent simple interest against
land held as collateral. But more than its low rates, what attracted
small farmers and craftsmen to the idea of the old colonial bank was
the Loan Office's relatively democratic credit policies. Unlike the
Bank of North America, which made loans only to the secure and

wealthy, the Loan Office offered credit to any property-holder in amounts up to £100. In its 32 years of existence, the Loan Office made most of its loans in small and moderate amounts to family farmers (67%) and craftsmen (25%) throughout the province. Especially favored among craftsmen were tailors, weavers, blacksmiths, carpenters, and cordwainers, each of whom accounted for 2 to 3 percent of all loans.[63] The Loan Office was replaced by direct English credit after 1755, but its principles were not forgotten. In 1786 John Smilie, the Constitutionalist's foremost rural spokesman, gestured toward the Loan Office when he defined the bank issue as a question not of "whether we shall have paper money—but who shall omit it—the bank or the state?"[64] Later the same year, "A Fellow Citizen" made the point even clearer, declaring that "no system can be devised equal to a Loan Office, established on a permanent footing."[65]

In the face of widespread discontent among the middling and laboring classes, the Constitutionalists used their slim majority in the 1784 assembly to work toward the dissolution of the bank. The Report of the Assembly Committee, issued the following spring, made the Constitutionalist position clear. Calling the bank an "enormous engine of power" the report argued that it "would be totally destructive of that equality which ought to prevail in a Republic."[66] Invoking the language of the 1776 Declaration of Rights, the committee claimed that "the accumulation of enormous wealth in the hands of a society, who claim perpetual duration, will necessarily produce a degree of influence and power which [can only] endanger the public safety."[67] In the end, they warned, "we see nothing which . . . can prevent the directors of the bank from governing Pennsylvania."[68] This, of course, was precisely the Republicans' intent. On March 26, 1785, the Constitutionalists entered a bill to repeal the original act of incorporation.

The bank lost its charter in 1785, and for the next two years Morris and the Republicans mounted a determined campaign for its rechartering. Pro-bank articles became a staple of the Philadelphia press as its supporters appealed to the public in conciliatory tones. One, "An Old Banker," tried to convince skeptics that "the Bank operates as a great machine, uniting and combining the whole power of the com-munity into one strong, forcible movement."[69] And Paine, who had been commissioned by Morris and the Republicans, wrote that he had "always considered the bank as one of the best institutions that could be devised to promote the commerce and agriculture of the country . . . I have always been a friend to it."[70]

The election of October 1786 brought the Republicans back into control of the legislature. Thereafter the rechartering of the bank was assured. Meeting in March 1787, the assembly promptly rechartered the bank, but not without a series of restrictions placed on its opera-tions by assembly Constitutionalists. With the new charter secured, the Republicans could now turn their full attention from national and state affairs to recapture the magistracy of Philadelphia.

The climax of the Restoration was a local affair. Driven from municipal power during the Revolution, the city's old-line families had long sought to end assembly control over city affairs and return it into the hands of Philadelphia's pre-Revolutionary establishment. Their chosen vehicle was a latter-day revival of the Corporation which had ruled Philadelphia from 1701 until its overthrow in 1776.[71] From its inception, the colonial Corporation had been an anachronism, drawing its prototype from the administrative appa-ratus of medieval English towns. In authorizing such an outmoded form of civic organization, William Penn had hoped to bind his colony's first families to his proprietary interests, and the Corporation quickly evolved into a closed, quasi-aristocratic body in which offices were handed down from father to son or shared within the close, intimate world of the city's wealthiest families.

In its actual functioning the colonial Corporation had worked less as an administrative body than an elite social club. The members of Philadelphia's Corporation early handed their administrative duties over to subordinate boards and commissions and transformed the Corporation into a brokerage house specializing in the circulation of wealth, family connections, and prestige.[72] But, like all patrician institutions, the Corporation had a public function as well, serving as one of the primary agents in the negotiation of deference and the bolstering of established authority. From its inception the Corporation had adopted a pompous and stylized public ceremonial designed to overawe the "lower orders" of the city. A visitor to the Quaker City in 1744 described one of these ceremonies, the procession preceding the public reading of the Declaration of War against France. The procession moved with great solemnity through the city with the sheriff and his constables acting as ushers ritually clearing the streets and purifying the path for the march of the notables. The procession was led by "his Honor the Governor, with the Mayor on his Right, and the Recorder of the City on his left hand."[73] Following closely behind were the remaining members of the Corporation. In the enactment of this civic ritual the social hierarchy of Philadelphia was not only given a visible and dramatic form in the order of march, but the unity, strength, and magnificence of the city's establishment was powerfully displayed as a check upon any latent plebeian opposition to patrician rule.

The Corporation itself collapsed in the revolutionary upheavals of 1776, but the social relationships which both expressed and sustained it continued throughout the war as the Powels, Norrisses, Shippens, and Willingses waited upon more auspicious times. The Republican resurgence of the 1780s gave to Philadelphia's establishment the political climate as well as the means to complete their restoration. Beginning in 1781 with a petition calling upon the assembly to restore the Corporation, Samuel Powel, Matthew Clarkson, John Wharton, and Miers Fisher—the leaders of established Philadelphia families—worked with increasing success to return the city's pre-Revolutionary oligarchy to power.[74] Finally in March 1789, with the Republicans in firm control of the assembly, the federal Constitution ratified, and the Restoration secure, the assembly performed its final act of retrenchment and returned the reins of city government into the hands of Philadelphia's first families.

Unlike its predecessor, the post-Revolutionary Corporation was more than a marketplace of ritual deference.[75] In the aftermath of intense popular mobilization during the war, Philadelphia's men of wealth could no longer depend upon the automatic allegiance of large numbers of the middling and laboring classes. The new city charter reflected the insecurity of restored power and the uneasiness with which its holders confronted an informed and enfranchised populace. Patterned on the new federal and state constitutions, the city charter of 1789 provided for two ranks of officials: aldermen whose terms ran for seven years, and common councilmen whose terms ran for three. The charter also provided aldermen with judicial functions, making them *ex officio* justices of the peace and creating a mayor's court in which they were the sole judges. It was also the aldermen's singular privilege to elect Philadelphia's mayor from among their number.

The most remarkable feature of the charter, however, was neither its division of powers nor its concentration of power in the hands of the aldermen; rather it was the provision that electors for aldermen be drawn solely from the city's freeholders. This provision virtually excluded journeymen, laborers, and mariners, as well as many small, independent craftsmen and retailers, from the local suffrage. The charter had accomplished in Philadelphia what the most aristocratic Republicans could never have hoped to achieve in Pennsylvania on the whole: the disfranchisement of those classes who had driven them from power in 1776. The passage of the charter by the state legislature in a state with virtual manhood suffrage is testimony both to the disorganized state of political opposition and to the weakness of Philadelphia's popular movement at the advent of the new federal, state, and municipal governments.

The first city elections under the new charter were held in March 1789, and laboring-class Philadelphians found Powel and Clarkson sitting as aldermen and Wharton and Fisher occupying common council seats. Turning to the task of filling the office of mayor, the aldermen unanimously elected Samuel Powel, the last colonial mayor of Philadelphia, to his old seat. No more apposite symbol of the successful restoration could have been devised.[76]

The revival of the colonial corporation, the rechartering of the Bank of North America, the nationalization of commercial regulation, and the repeal of the Constitution of 1776 were the major programs of the Republican Restoration. Once they were in place the rich and well-born were assured a dominant role in the functioning of the economy and were also assured a hegemonic position in state politics and in society as a whole. In less than a generation Revolutionary hopes for a republic of small producers had been defeated by merchants and speculators in land, currency, and human needs. It remained for the working people of Philadelphia to wring a victory from this defeat.

II

The Restoration was the background against which Philadelphia artisans sorted out their lives, argued about issues, and thought in new and creative ways about their interests. After the disruptions of the Revolution, there were lines of custom to rearrange, new and more numerous goods to produce, and changing market relationships to adjust to. As the 1780s wore on, there were new concerns as well: rising prices of food and materials; the lack of paper currency; vendue sales of cheap and shoddy British manufactures; and the cycle of debt, failure, and debtors' prison.

It would be uncharitable, in the context of a chaotic economy sliding from wartime boom to postwar depression, to expect more of working people than a shunting between anxiety and discouragement as they fought to keep themselves employed and their families clothed and fed. Certainly, the poorhouse dockets and the psychiatric notebooks of Samuel Coates record some of the personal failures.[77] And yet we find in the 1780s not only a remarkable resiliency on the part of working people but an even more remarkable, and important, forward motion.

Because the Restoration was carried out in a context of manhood suffrage and active political campaigning, it has been conventional for historians to assume that the working people of Philadelphia allied with the party closest to their interests. In the dominant view, the popular radicalism of the Revolution declined in intensity by war's end, the people became more sensible, and in the 1780s their fervor was disciplined and channeled into the disputes of party and faction.[78] It is a short distance from this view to a general turning away from consideration of the popular-radical tradition itself into a discussion of parliamentary maneuverings in which the stratagems of legislation are substituted for the thoughts and actions of the people themselves. In this shift in emphasis the ideas and organizations of artisans and ordinary working people are marginalized, if not hustled from the stage altogether. The members of the Revolutionary committees, the Committee of Privates, or the militia at Fort Wilson are remembered only inasmuch as they held office or participated in government. Noting that Peale and Hutchinson suffered defeat at the hands of resurgent Republicans in 1780, that Matlack was driven from his post as Secretary of the Supreme Executive Council by a Republican-dominated assembly in 1782, that Rush became a convert to the Republicans, and that Paine was plumping for Morris's bank and publicly arguing that unicameralism did, after all, allow "too much precipitancy" in legislation, historians have allowed the popular movement to follow its radical leaders into an oblivion of neglect.[79]

The elisions of this parliamentary perspective are substantial. Not only are the lives and experiences of a large majority of the city's population negated, but the ways in which working men and women activated their own sense of justice and took action to ensure that equality would be a social practice and not a mere rhetorical staple are foreclosed. There is condescension in this, and also a smugness which celebrates mechanical democracy while ignoring its real functioning in everyday life. The tragedy of this view is that it is as blind to the workings of democracy in the shops, neighborhoods, and organizations of common men and women as it is silent about the ways in which ordinary people have tested their ideas about democracy against the burden of their experiences and in the process have sought to change their world to meet the standards of their judgment.

The breakup of the radical-popular movement in late 1779 made the political allegiance of craftsmen and other workingmen problematic for the leaders of the Republican and Constitutionalist societies. But no less so for historians. Secret ballots and the lack of residential segregation in the city make it impossible to trace voting patterns according to even the broadest of class lines.[80] Nor did any laboring-class organization of the 1780s endorse one party over the other. The Republicans often assumed that the middle and lower ranks of craftsmen sided with the Constitutionalists, but this was more the result of the exclusion of ordinary artisans from Republican party counsels than a true account of the situation.[81]

In the end, however, the question of laboring-class party preference is of less importance than the fact that their preference was expressed at all. Philadelphia's working-class movement was the first, and in the light of subsequent history, one of the few, to have at its inception direct access to the ballot box. This was profoundly important, for it gave to carpenters, tanners, cabinet makers, and draymen—the whole gamut of skill and occupation—a political presence which in England had to be gained against determined and powerful repression. Unlike the English working class, which had to forge its collective identity in a context of severe repression and episodic protest, the practice of representative democracy provided Philadelphia's laboring classes with a legitimate focal point for discussion and debate.[82] This early inclusion within the pale of formal politics meant that Philadelphia workingmen could give their minds more fully to developing their ideas, their community, and their moral judgments—in short, to articulating their own way of life.

It is possible to trace a laboring-class political presence after the demise of the radical-popular movement in 1779. Six months after Fort Wilson Philadelphians awoke to find a hastily printed broadside posted on doors and lampposts throughout the city. Written by members of the militia who identified themselves as part of the "laboring poor," the broadside called for a public meeting to address mounting laboring-class concerns about high prices of the "necessaries of life" and continuing inequalities in the state's militia law. Given the alignment of forces following Fort Wilson, the militiamen were aware of the likelihood of civil repression (indeed the Light Horse patrolled the city on the day scheduled for the meeting) but declared that they were nevertheless "determined to be free."[83] The proposed meeting never took place, but the determination of workingmen continued. In this the signature adopted by the militiamen was prophetic; if their object remained the same, their methods would now be "slow and sure."

As this non-incident indicated, artisan discontent did not so much diminish in the early 1780s as change its form. Discontent might still be expressed in a direct way—the harassment of wealthy Quakers following the American victory at Yorktown and the refusal of the militia to muster against mutineers in 1783 are two examples—but after 1780 we find increasing evidence of self-organization among city artisans coupled with a rejection of middle-class, articulate leadership. By the close of the Revolution it had become clear to journeymen and small independent producers that the community of interests of a small producer's republic would be brought about only by their own efforts, both in their shops and in the political arena as well.

The success of laboring-class self-organization can be measured indirectly by the continuous stream of propaganda directed at workingmen on the issue of import tariffs, beginning in 1783. For while the opening of American ports to British manufactures in that year had been a boon to Philadelphia's import merchants, it was a devastating blow to most of the city's mechanics whose higher costs of production made them unable to compete with the low-priced fruits of English industrial development.

In June 1783 "A Friend to Mechanics" asked the artisan readers of the *Pennsylvania Gazetteer* to consider whether "those industrious citizens, who have so capitally contributed to the independence of their country, [shall] now be forgotten, and with their families become the objects of the greatest distress, for want of employment in their respective branches."[84] The question had but one answer and the writer urged craftsmen to muster their collective political power and pen "a spirited, but decent remonstrance to your representatives . . . for a redress of grievances."[85]

Two weeks later, on July 6, a "number of Mechanics and Manufacturers" followed this advice and held a general meeting to discuss "business of the utmost importance to themselves, to their families, and to their country."[86] The result was a mechanics' memorial that called upon the Assembly to restrict or prohibit the importation of English goods into the state.

The letter from the mechanics' "Friend" and the July 6 meeting were small but important events in the history of Philadelphia's working class. The June letter was the first reference in the Philadelphia press to the mechanics as a group distinct from "the people" at large. It was also a recognition that the resumption of trade with Britain had set the interests of the maritime trades, textile workers, and iron workers against those of the importing merchants of the city.

The mechanics' memorial brought forth no action from the assembly, and in October a correspondent to the *Pennsylvania Gazette* summed up the situation in Philadelphia in heightened class terms, describing "rich and oppressive monopolizers, oppressed and starving manufacturers; tradesmen and mechanics slighted and trampled on; freedom and patriotism expiring."[87] Conditions in the city were clearly worsening, and the *Gazette* issued a warning to city workingmen: "Mechanics . . . beware! The enemies to freedom, to virtue, to trade, and the community, cannot be friends to you."[88]

These episodes of the summer and fall of 1783 portray a hyperbolic and self-interested campaign to woo the laboring classes to the side of the Republicans. They were, in part, obviously this; but only in part. They were also a recognition that some form of organization existed among city artisans. The call for a remonstrance in June implies a working committee to draft the petition and suggests some organized means for its ratification. The call for a general meeting of mechanics may have been addressed to informal trade organizations as well as individuals. The same may be said for "A Friend to All Mechanics," who advised "if you gain any interest from persuading the skunks [Constitutionalists] to go with, join and vote for your men, so as to become pliant to and execute your own views, you are so far right; it is justifiable. But be guarded against any further political intimacies or connexions."[89] The tone here suggests the second person plural and again indicates some degree of organization among Philadelphia workingmen. Certainly such informal groups existed among eighteenth-century English textile workers (many of whom migrated to Philadelphia after the Revolution) and also in many other trades where production was a collective, social undertaking.[90]

Looking beyond organization, these episodes also indicate an artisan outlook that would last well into the nineteenth century: the ruling class seen as "rich and oppressive monopolizers"; the dignity of workingmen "slighted and trampled on"; "freedom and patriotism . . . virtue and community" expiring in the face of a self-interested ruling class. The thoughts run back to the English revolution and the rights of small producers; and they run forward to the Workingmen's movement of the 1820s. The admonition given in 1783 to "remember that the contention between the two parties which divide and distract the state, is for POWER only"[91] is the same given in 1647: "What else but your Ambition and Faction continue our Distractions and Oppressions?";[92] and given again in 1828: "The majority of the working men . . . having mainly contributed to the elevation of their ambitious favorites are doomed to sink again into their former insignificance."[93]

Disillusionment with the political process does not lead to apathy or reaction in any automatic way, however. The record of political disappointment and tepid leadership experienced by Philadelphia's laboring classes during the war years was met not with resignation but with a renewed sense of purpose and collective mission. One aspect of this renewed determination was the flowering of journeymen's trade societies during the 1780s. These societies met to settle the annual price of work between journeymen and employing masters, but they also provided a forum for discussions of the well-being of the trade, as well as about larger political and social issues. These organizations were often informal, unincorporated, and lacked official recognition, but in bringing together the workers in a shop, or the journeymen of an entire trade, the journeyman's societies created their own legitimacy of numbers that would lead to the creation of more formal organizations as events, and a rapidly changing economy, required.

It is difficult to account for the influence of these early trade societies with any certainty because of their general lack of record-keeping. This is especially true of the informal societies which may not have kept records, but is also true of the organized societies for which not even minutes survive. It is nevertheless unlikely, given what we know about similar societies in England and subsequent societies in America, that their discussions were limited to "bread-

and-butter" issues. With a modicum of imagination, we can link together the political lessons learned during the Revolution with these trade discussions and find as a result an altogether novel program for laboring-class political action.

This program was first adumbrated in October 1783 by "A Brother Mechanic." In an election address to the mechanics of Philadelphia, he alluded to the predominance of the laboring classes in the Revolutionary armies and to the sacrifices which had made their "right to a full representation in the House of Assembly incontestable."[94] He went on to remind his fellow mechanics that "you will . . . have it in your power to redress yourselves" in the coming election "by forming and supporting your own ticket independent of party, the bane of society."[95] Here were the methods of the radical committees of 1775 and 1776 brought back in a new context. In the face of the Assembly's failure to act on their July memorial, many among the city's laboring classes were now calling for an independent artisan politics that would avoid the self-interestedness of existing parties, would counteract the rural domination of the legislature, and would lead to the elimination of ruinous British competition.

"Brother Mechanic"'s address opens an important window on laboring-class thought and temper. Faced with a farmer-dominated legislature that saw in tariffs only higher consumer prices, by October 1783, neither party had acted on the mechanics' memorial, leaving Philadelphia's craftsmen at the mercy of British imports and a free market. The conclusion to be drawn was obvious: as one city craftsman put it, both the Republicans and the Constitutionalists were interested "in POWER only" and if "fixed in the saddle of State . . . [would] consider [mechanics] as a useless set of beings, who ought to be fed with 'butter-milk and potatoes,' while the darling objects of their attention and pursuit ate the 'flesh-pots and onions.'"[96] The antipathy toward wealth and the ridicule of social inequality and ruling-class pretension voiced here were already staples of laboring-class thought by the mid-1780s. What was new was the attitude toward politics. The unresponsiveness of both parties to the craftsmen's appeals merely confirmed the lessons learned in 1779; the only way to preserve their values and livelihoods was, as

"Brother Mechanic" declared, to "stand on your own ground" and "be not diverted or deceived by either party—make no distinction, but continue in the line you have chalked out for youselves."[97]

"Brother Mechanic" ended his piece with a concrete program for redressing laboring-class grievances. In an age when nominations were handed down and not upward, he advised his fellow craftsmen to "nominate your own Representatives, and if you fail in your choice, establish a Society of Mechanics, despise and reject party interests, and repeat the experiment at the next opportunity."[98]

We do not know whether the workingmen of Philadelphia took an independent course in the October 1783 election, but it is likely that they did not, at least in any organized way. "Brother Mechanic"'s address was printed only the day before the election, hardly enough time to nominate representatives or to draw up an election ticket. But his call did not go unheeded: cordwainers continued to discuss Morris's bank; carpenters and sawyers joked about the rich man whose house they were building; taverns, workshops, houses, and the streets themselves continued to be meeting places for political discussion. The call would need only time to be answered.

By the election of 1784 the unending stream of cheap British manufactured goods had ruined the competencies of growing numbers of Philadelphia craftsmen and made the importation of British goods the focal point of laboring-class discontent. Over a signature recalling the seventeenth-century movement that had championed free will and individual religious judgment over the authority of Calvinist orthodoxy, "Arminius" addressed the mechanics of Philadelphia and again urged them to adopt an independent political stance. "Amid the views of ambitious leaders and cunning partizans the good of the public is lost and forgot. The struggle is for faction, not the interest of the state. What then remains for a freeman to do? He should certainly assert his dignity and consequence like a man in the scale of things and great theatre of action."[99] "Arminius" went on to associate independent political action with independent religious beliefs, and then continued, "the mechanics form the bulk of the city—and the sin is at their door, if they do not resolutely league together, abandon Constitutionalists and Republicans, and in the

97

virtuous struggle, endeavor to defend themselves, their families, and their country."[100]

"Arminius"'s ideas were reinforced in the same issue of the *Pennsylvania Gazette* by the thoughts of "An Old Mechanic":

> The mechanics of this city have as good a right to participate in the benefits of Independence as any class of men whatever—and yet they have been neglected even by their own representatives. It is therefore most devoutly to be wished that you will lay aside all party disputes and animosities which have but too fatally divided your interests . . . the true interests of your country, the welfare of yourselves, and the preservation of your families, require and demand your unanimity in the appointment of independent, good men to represent and legislate for you.[101]

In the arguments of these anonymous artisans we find clear evidence of an artisan political presence in Philadelphia during the early postwar years. After 1783 Republicans and Constitutionalists gradually understood that workingmen were a sufficiently organized, conscious, and determined class that their particular needs and interests had to be taken into account. The unexpected strength of this laboring-class presence can be measured by the degree to which the established parties followed, rather than led, laboring-class actions. Spurned by two successive legislatures, Philadelphia workingmen turned their backs on the Assembly and attempted to take matters into their own hands. On March 10, 1785, Philadelphia's master cordwainers met and resolved not to "buy or sell . . . nor mend nor suffer to be mended by any in our employ" imported boots and shoes.[102] Richard Collier, the clerk of the master cordwainers, also reported a similar resolution coming from the independently organized journeymen of the trade. What they could not accomplish through electoral channels, craftsmen would now accomplish through the economic pressure of the boycott; Philadelphia workingmen were lurching away from the political orbit altogether.

Frightened by the assertive independence of city craftsmen and anxious about the extra-parliamentary direction they were taking, Philadelphia's Merchants' Committee met on June 2 and issued a public call for a meeting to discuss non-importation. The press had been active on the issue for over a month, reprinting letters from Boston pitched at that city's laboring classes. Significantly, two of the reprinted letters came from a descendant of Boston's founding family, John Winthrop, Jr., whose moniker, Joyce, Junior, recalled Cornet George Joyce, the humble tailor who arrested Charles I in 1647 and placed him under the control of the rank and file of the New Model Army.[103]

> Rouse. Rouse then my countrymen! extirpate the viper that is gnawing our vitals, and which threaten speedily our dissolution . . . the same spirit which we exerted before the year 1775, is yet latent in our business. JOYCE, Jun.[104]

> Let us form ourselves into a body as ready and firm as in former times, when the interest of the community was not more in danger than it is now. JOYCE, Jun. [105]

A third letter sounded a dramatic resonance with the events of Fort Wilson and the desertion of the laboring-class militia by the radicals:

> Though timid Whigs and cringing panders may cry no mobs and riots . . . be assured that the voice of the people is the voice of God. LIBERTY[106]

Finally aroused by the growing anti-parliamentarianism of the laboring classes as well as the weighty concern of the Merchants' Committee and the lobbying of the Philadelphia press, the Constitutionalist-dominated assembly sought to repair electoral channels, and to ensure their own re-election, by passing a weak tariff bill two weeks before the October 1785 election. The passage of the tariff averted an immediate political crisis but did not solve the problem of English competition for Philadelphia's laboring classes. By 1787 they would follow their collective interests and support the federal Constitution, which promised full and effective tariffs on a national level.

The town meeting held on June 20 tells a great deal about class relations in postwar Philadelphia.[107] For the first time since the town meeting of September 1779 which opened the issue of price regulation to public debate, the working classes were asked for public consent outside of the polling place on an issue vital to their welfare. The Merchants' Committee had called the meeting to organize public approval for their plan to nationalize commercial regulation. They easily received general approval for their call to an end to undercutting imports, but when they formed a committee to memorialize Congress, it was composed solely of merchants. Someone called out from the crowd that mechanics had as much at stake in the matter as the merchants and they ought to be represented on the committee. The crowd agreed, but when the list of committeemen was published, the merchants betrayed their true feelings and fears concerning the city's artisans and workingmen. The seven "mechanics" added to the original committee of twelve merchants bore little resemblance to the authentic mechanics' committees of the Revolutionary era. The mean wealth of the "mechanics," as they were assessed in 1782, was £1,314, little more than 10 percent below that of the merchants themselves.[108] That the "mechanics'" committee was not made up of craftsmen is apparent, not only from their wealth but from the social position of some of its members. Anthony Cuthbert was the son of Philadelphia's wealthiest shipbuilder, and Joseph Marsh's claim to the title of mechanic derived solely from his ownership of Philadelphia's famous Globe Glass Works. Judging from their high assessed wealth, the "mechanics" were in truth employers and not mechanics in the usual sense of the term. As the deceptive composition of the Merchants' and Mechanics' committees made clear, the working classes of Philadelphia were important enough to be courted yet strong enough to be feared.

A month after the merchants' clumsy attempt to control artisan politics, an anonymous writer responded to the duplicity of the Merchants' Committee with an appeal for artisan political action. Addressing Philadelphia workingmen in the pages of the *Pennsylvania Gazette*, the unsigned author called on "the various bodies of artisans and manufacturers in this city to assemble immediately, and form committees of their most intelligent members, for the purpose of drawing up plans of the duties necessary to be laid on those European articles which rival the establishment of manufactures in this state."[109] In this letter was confirmation not only of the widespread existence of trade organizations in Philadelphia but also of the workingmen's insistence upon charting an independent course.

The logical extension of the activities of independent trade societies has always been a recognition by their members of their common interests as working people and a realization of the power gained by joint action in a general trades and political union. This step was not taken by Philadelphia artisans until the late 1820s, but the point was made as early as September 1785, when the following proposal appeared in the *Pennsylvania Gazette:* "We would recommend an association of your tradesmen and manufacturers formed upon the most extensive basis, and supported upon the most liberal principles; we may then hope the manufacturers of this country will flourish when each man becomes interested not only in his own branch, but in those of his brethren."[110] The creation of just such an association of workingmen "upon the most extensive basis" was a project that would occupy Philadelphia workingmen for the next forty years.

By laying the Republican Restoration and laboring-class independence side by side across the 1780s, we can see the artisan political crisis for what it was. Far from capitulating to the restoration of old-line wealth and power or a party system unresponsive to their needs, Philadelphia artisans began mapping out a political and economic plan for a society of small producers: a society in which independent politics and craft organization would inaugurate a world of honorable labor and harmonious independence for the producers of America.

CHAPTER 4

Hegemony and Counter-Hegemony: The Rebirth of Popular Radicalism, 1790–1795

The success of the Restoration in Philadelphia, combined with memories of the betrayals of 1779 and the inability of Constitutionalists to accommodate their pro-country policies to the needs of urban craftsmen demonstrated the futility of party politics to many of Philadelphia's workingmen. There was an unmistakable tone of disenchantment and disgust in the post-ratification air. Freneau's *Independent Gazetteer* spoke in 1789 of the new catechism of the times which elevated wealth over all other values:

What is the chief end of Man?
To gather up riches—to cheat all he can,
To flatter the rich—the poor to despise,
To pamper the fool—to humble the wise.

The rich to assist—to do all in his power
To kick the unfortunate still a peg lower.

To deal fair with all men, where riches attend them,
To grind down the poor, where there's none to
defend them.[1]

During the summer of 1789, as many celebrated the establishment of the new national government, "A Plebeian" wrote of the evils of "fat offices and lucrative offices." "The peace of the community is destroyed by furious squabbles for them," he contended, and "we are split into parties, and set to quarreling and abusing one another, that we may decide who shall have them."[2] This sordid state of affairs would not have occurred, he argued, under the democratic Constitution of 1776, "that system which has been so much abused and which, at present is so much hated by our great men and patriots."[3]

We can hear echoes of the radical Whig tradition here: the contempt for placemen and the perversion of good government by cash, honors, and privilege. Yet there is more than that. The basis of argument is not in natural law but in the well-being of the community. It is the "peace of the community" that is destroyed and not the proper balance of government. "A Plebeian" celebrated the Constitution of 1776 not for its balanced powers, but because it was a more direct and democratic instrument for the expression of community sentiment.[4] Not least, there was the biting irony used to condemn the "great men and patriots" who had abandoned Revolutionary principles for a modern code of private remuneration and public self-advertisement.

A fortnight later, "A Plebeian" continued his civic jeremiad by drawing a sharp line between political principle and the pervasive venality of the times. Surveying the events of the past decade, he concluded with ill-concealed sarcasm that "Locke, Sidney, Harrington, and Montesquieu were all mistaken in supposing that government was created for the good of the people. The truth is, that government was created for the purpose of maintaining men who want talents or industry to maintain themselves."[5] The craftsmen who had watched their Revolutionary leaders carve out lucrative livelihoods for themselves in postwar government service might understandably have muttered "Amen," pulled their caps down more tightly over their heads, and, with a confirmed cynicism, turned their attention to their own, somewhat smaller, fortunes.

But the workingmen of Philadelphia had been disillusioned before, and each time had found a source of strength and resiliency in their collective resources and in the small producers' creed. They would draw upon these sources again in the 1790s as they faced the consolidation of elite power and, with it, the emergence of an hegemonic liberalism that threatened to undo their whole way of life.

I

In the last decade of the eighteenth century, liberalism was much more than a set of abstract doctrines; it was an active body of ideas used to promote and legitimize the contemporary realities of economic and political change. Against the incipient populism of "Plebeian" and the small-producer tradition of Philadelphia artisans, liberalism envisioned a world with the stops pulled out, a world in which the reciprocity of craft and community would be supplanted by the unfettered pursuit of individual self-interest.

Some notion of the distance between this emergent liberalism and the residual traditions of popular radicalism can be gauged by comparing "Plebeian"'s views with that protean text of American liberalism, James Madison's tenth essay in the collected *Federalist Papers*. In it, Madison noted the same "prevailing and increasing distrust of public engagements" that "Plebeian" pointed to in 1789.[6] He also agreed with Philadelphia workingmen that this distrust was the result of "a factious spirit [that] has tainted our public administration."[7] The city's poorer craftsmen, journeymen, and day laborers would certainly have concurred, too, with Madison's judgment that "the most common and durable source of factions has been the various and unequal distribution of property," and that "those who hold and those who are without property have ever formed distinct interests in society."[8] This had been their experience since 1779.

But while Madison and Philadelphia's laboring classes might have agreed on these facts, they drew radically different conclusions from them. Madison's intent was to cloak the new government in a fine cloth of legitimacy, a cloth woven in equal parts of absolute property rights and an ultimate distrust of the people. In setting his loom in this way, Madison attacked the egalitarian and democratic tenets of the small-producer tradition. "Democracies," he asserted, "have ever been spectacles of turbulence and contention [and] have even been found incompatible with personal security or the rights of property."[9] In *his* federal republic, on the other hand, the wantonness of the people would not be allowed to rot the social fabric because "the public voice, pronounced by the representatives of the people, will be more consonant to the public good than if pronounced by the people themselves."[10] To Madison, ordinary people too often went beyond the pale of rationality and were often incapable of recognizing their own best interests; if given too loud a voice in government, they would most certainly end by shouting down civil rights and property rights altogether.

Madison, of course, had little direct knowledge of the thoughts of working people. If he had, he would have found that personal security and property rights were as important to them as they were to his Federalist allies. Workingmen had demonstrated their commitment to "liberty and property" in their service with the Revolutionary militia, and equally in the intricate definitions of property rights they had articulated during the price-control movement of 1779. Granted, theirs was a more complex and intensely social conception of the rights of property holders compared with the legalistic notions espoused by Madison and the early Federalists. But as the idea of competency confirmed, property had always played a central role in artisan notions of right. The supporters of the Constitution, on the other hand, like most men of wealth in early national America, viewed absolute property rights as fundamental, not only to their own security but to the continued health of the national community as a whole. As an early Federalist explained, "unrestrained liberty of the subject to hold *or dispose of* his property as he pleases, is absolutely necessary to the prosperity of every community, and to the happiness of all individuals who compose it."[11]

So far as personal property was concerned, the blacksmith or carpenter who owned his tools and a few household items would have agreed; there was every reason to make one's tools, personal and family effects, and, for the minority of craftsmen who did not rent their quarters, one's home secure. A different conception applied, however, when personal property rights were extended beyond the competence of the small producer to include large accumulations of wealth. It was this trespass of unconditional property rights into the established realm of community regulation that divided the craftsman's notion of property rights from that of his patrician counterpart.

Perhaps more than any other of his arguments, Madison's claim that "the latent causes of faction are thus sown in the nature of man"[12] set patrician culture against that of Philadelphia's laboring classes. Following a line of argument that remains dominant to this day, Madison located the cause of differences in wealth and social position in a "diversity in the faculties of men," from which "unequal faculties of acquiring property" directly followed.[13] In the developing logic of post-Revolutionary liberalism, such reasoning would lead directly to a justification of social, political, and economic inequality that was repugnant to the craftsman's deep-rooted democratic traditions. In Madison's hands, that logic had already become the basis for a denunciation of such "improper or wicked project[s]" as paper money, abatement of wartime debts, and a more equal division of property, all among contemporary popular-democratic demands.[14] In the end, Madison denied the whole logic of the small-producer tradition when he wrote that "neither moral nor religious motives can be relied on as an adequate control [of individual] passions and interests."[15] The quest for moral justice was, as Madison was later to discover, the driving force of the laboring-class movement, and it was upon just such moral controls that the small producer vision of a just and equitable society rested.

II

The late 1780s and early 1790s marked a period of intellectual crisis for Philadelphia's laboring classes and for the small producer tradition that informed their everyday lives. The conservative resurgence that followed on the heels of the ratification of the federal Constitution in 1788 not only brought a political and social restoration reminiscent of late seventeenth-century England but produced a series of attacks on artisan moral and intellectual traditions, of which the tenth Federalist was only one among many examples. Patrician Philadelphians were firmly in control for the first time since the 1760s, and they labored to consolidate their rule by establishing dominion over intellectual as well as social and political life.

Once in office, the members of Philadelphia's reinstated Corporation moved rapidly to assert their authority and to reorder civic life in conformity with elite tastes and interests. Surveying the effects of the first three years of the Corporation's existence, "A Drayman" recalled that the Corporation had "enacted [an] abundance of Ordinances, said to be intended for the better regulating . . . the police of the town," but which had in fact "neglected to hinder the rich from oppressing the poor."[16] As proof he appended a lengthy list of grievances which ranged from the "driving off [of] hucksters and retailers, both young and old from the marketplace," to the Corporation's concern that "printed pounds of butter, and fair loaf bread, be full in weight and measure, so that the *well-born* may be enabled to eat them both cheap and in taste, while the rendered bit of lard, and humble buckwheat cake, a material part of the laboring poor's food, is left to the mercy of every miscreant huckster, to sell as they please, whether light or heavy, good or bad."[17] Not content with ensuring the elite's unencumbered passage through Philadelphia's streets and the purity and full value of their food, the Corporation sought in 1795 to protect their homes from the threat of fire, and incidentally to lower their own insurance rates, by prohibiting the building of wooden houses in the city. This seemingly prudent piece of legislation in fact cut deeply into the competency and lifestyle of city craftsmen. Then as now, brick houses were expensive and artisans and ordinary workingmen depended upon the low rents that frame houses afforded. Moreover, craftsmen had long depended upon their own labor and the easy credit that could be obtained for wood and other construction supplies to build their own homes within the city limits. In banning affordable housing, the Corporation seemed bent on the total domination of Philadelphia's social, political, and economic life, even if, as one workingman concluded, the result was "to drive industrious and poor mechanics and tradesmen from the city."[18]

The dominion that Philadelphia patricians were fast establishing over city life gained added strength from the presence there of both the state and national governments throughout the 1790s. Living at the hub of American politics, Philadelphia Federalists could mingle freely with state and national elites, absorb their political wisdom, and bask in the reflected light of their prominence. The elaborate spectacle that accompanied every public event, from Washington's ostentatious arrival in the city to the opening of the federal Congress, served to reinforce the magnificence of elite rule. Even the city's architecture echoed the hegemonic powers of a confident ruling class as a spate of new public buildings appeared almost overnight, all constructed in the latest "Federal" style that aped contemporary English aristocratic taste. Not to be outdone, Philadelphia's wealthiest families sought to establish their own claim to architectural and cultural prominence by building new estates in and around the city, the most lavish of which—William Bingham's urban mansion—occupied the better part of a city block. Bingham had taken his plan from the Duke of Manchester's London townhouse and in its replica Anne Willing Bingham presided over a salon which was the rival of any in contemporary Paris.[19]

But even more important than country estates or lavish entertainments in creating an aristocratic culture in Philadelphia was the presence of Washington in the city. His frigid and detached demeanor and his decorous afternoon dinners, described by one guest as being "as grave as at a funeral," set the tone of the federal administration in Philadelphia and set the standard for the city's elite as well.[20] While serving as President, Washington not only acted the part of an aloof monarch, but he drafted an elaborate formal etiquette to be followed by those in attendance at what quickly became known as his "republican court." Absorbing the style of federal rule, Philadelphia's elite emerged from the 1790s with a reputation for haughtiness unmatched by any other urban ruling class in the new nation.

Hegemony, though, is not built of wealth, parliamentary dominance, and state spectacle alone. One of the essential tasks facing any ruling class is to devise some means of making the unequal relationships of wealth and poverty, power and subjugation, privilege and impediment—the whole apparatus of class dominance—appear natural and God-given and thus legitimate in the eyes of the lower classes. It was to this task that Philadelphia patricians turned with especial vigor after 1790.

The dominant culture of Philadelphia Federalism spoke in several voices, but one of the most revealing was that of Pelatiah Webster, a wealthy Philadelphia merchant whose *Political Essays* appeared in 1791. Webster's views are especially instructive in a discussion of hegemony in the Quaker City because his *Political Essays* were the first systematic attempt to articulate a political economy which placed a mercantile class at the center of its analysis.

Befitting an economy in which wealth and social standing came from trade and finance rather than production, Webster placed the merchant at the hub of economic relations. Merchants, he wrote, were those "whose business of life, and whose full and extensive intelligence, foreign and domestic, naturally make them more perfectly acquainted with the sources of our wealth."[21] In this he was following in the train of his teacher John Witherspoon, Presbyterian president of the College of New Jersey and one of the new nation's foremost proponents of the Scottish Enlightenment. In his *Lectures on Moral Philosophy*, Witherspoon, too, had argued for the social necessity of a merchant class because, he wrote, "the great and sudden fortunes accumulated by trade cause a rotation of property" and prevent the formation of an otherwise inevitable agrarian oligarchy.[22]

In Webster's mind, merchanting was a calling that bestowed a special authority on its practitioners. Trade, he maintained, was "of such essential importance to our interests . . . that no sources of our wealth can flourish, and operate to the general benefit of the community, *without it*."[23] From this claim, Webster drew the conclusion that the well-being of any community rested upon the well-being of its merchants because, as he put it, the merchant's "particular interests are more intimately and necessarily connected with the general prosperity of the country, than any other order of men."[24]

In Webster's *Essays* we find the foundations of an argument for the legitimate dominance of a merchant capitalist class. From his vantage point merchants were "the natural negotiators of the wealth of the country" and thus "underst[ood] the interests and resources of their country, the best of any men in it."[25] After all, Webster pointed out, "the States of V*enice* and *Holland* have ever been governed by merchants [and] no states have been better served, as appears by their great success, the ease and happiness of their citizens, as well as the strength and riches of their Commonwealths."[26] To ensure a similar form of governance in America, Webster proposed that merchants from each state send delegates to Congress where they would form a "Chamber of Commerce" whose *"advice . . . be demanded and admitted* concerning all bills" dealing with trade.[27] Since there were few things which, in his view, did not concern trade, Webster sought by this device to establish a permanent and powerful mercantile voice in all national legislation.

Rule by merchant capitalists was, however, only one part of Webster's project. Throughout his small book Webster was equally concerned to establish the hegemony of capitalist economic thought.[28] His immediate concern, especially evident in his seven "Essays on Free Trade and Finance," was to provide for government finance on a free-trade foundation. "In fine," he wrote, "my great object is to get our *revenue fixed* on a sure and sufficient foundation, and our *expenditures reduced within the bounds of use.*"[29] To accomplish this Webster proposed a national budget that balanced government expenditures against immediate tax revenues. He strongly opposed foreign borrowing and debt financing based upon future taxation in favor of taxes on consumption. Such current taxes, he claimed, might hurt the taxpayer, but they would succeed in paying the costs of government as well as securing its credit and fixing the value of its currency.

The regressive nature of his fiscal schemes, which would have taken proportionately more from the pockets of farmers and craftsmen than from the purses of merchants, speculators, and large landholders, was irrelevant to Webster. With a shrug of indifference, he wrote that "it is rare that the people refuse burdens or even grumble under them, when, by general conviction, they are necessary for the public good."[30] As events would soon prove, Webster's arrogant insensitivity to the plight of the nation's middling classes belied his claim that "no men are more conversant with the citizens, or more intimately connected with their interests, than the merchants."[31]

All of Webster's arguments were predicated by one, great operative principle: the natural power of supply and demand to set the price of goods. In much the same way that John Witherspoon had advised Washington against fixing the prices of military supplies during the Revolution because it obviated the consent of buyer and seller and ran counter to the operations of a free market, Webster argued against communal regulation of prices because it would violate the "natural course of things" and lead to the production of shoddy goods and a general breakdown of commodity circulation.[32]

This was not all. Webster went on to claim that unregulated commerce was consonant not only with the natural law of individual interest but with the well-being of the community itself. "Freedom of trade," he wrote, "is absolutely necessary to the prosperity of every community, and to the happiness of all individuals who compose it."[33] Pursuing this argument to a remarkable conclusion, Webster declared that free trade would reconcile the interests of merchants, farmers, craftsmen, and, one presumes, poor working people alike. In his mercantile world, "every man will go to market and return in *good humor* and full satisfaction, even though he may be disappointed of the high price he expected; because he has had the full change of the market [he] can blame nobody."[34] In this extravagant claim for the social utility of market mechanisms, Webster made class an irrelevance; in his entrepreneurial world all are brought to a level in a marketplace where human needs and desires are counted in pounds, shillings, and pence. Webster's prescription for human happiness was an equally easy one: "Let everyman make the most of his goods and in his own way, and then he will be satisfied."[35] With such home-grown utilitarianism available, there is little wonder that America's rulers never adopted Bentham's "beatific calculus."

Webster's *Essays* represented a thoroughgoing program for the hegemony of a merchant capitalist class: free trade set against the satisfaction of common needs, secure credit gained at the expense of equitable taxation, government by and for the benefit of the wealthy at the price of democracy. In the final analysis, Webster's arguments substituted cash for the more complex relationships of community and in the process reduced civil rights to the rights of contract. He made this remarkably clear in answering the Anti-Federalists, who had complained that civil rights were not enumerated in the new

Constitution. On the contrary, Webster countered, "the Constitution contains a declaration of many rights and very important ones, e.g., that people shall be obliged to *fullfill their contracts*, and not avoid them by *tenders* of anything less than the value stipulated."[36]

Webster's program for mercantile hegemony had many supporters, including important figures outside the merchant community. Joel Barlow, for example, held to a similar political economy, going so far in his *Advice to the Privileged Orders* to equate "natural society" with "a company of merchants."[37] Even Thomas Paine, who had helped to rouse Philadelphia artisans to the cause of revolution in 1776, added his strong, if equivocal voice to the merchant's cause, arguing that trade in the new nation "can neither be too large, too numerous, or too extensive."[38]

What brought these disparate voices together was a national debate about America's economic and social future that began during the Revolution and reached a climax only in the 1790s. The central issue in this debate was nothing less than the proper place of commerce and industrial development in American society. Advocates of unrestrained commerce, such as Webster and Paine, hoped to see America become "one great commercial republic."[39] Others such as Albert Gallatin, an Anti-Federalist congressman from western Pennsylvania, saw the future development of America differently. Against Webster's dream of capitalist development, Gallatin offered a vision of America as a small producers' republic, a nation of farmers, artisans, and small manufacturers extending indefinitely into the future. Echoing the arguments of the Committee of Thirteen in 1779, the Anti-Federalists viewed trade as a necessary but secondary social function and merchants as the economic servants of the producing members of society. Even manufacturing, while a productive activity, required subordination to the needs of the nation's small producers, Gallatin argued, lest America follow Britain's path toward industrialization and experience its attendant social and political upheavals. As George Logan, a Philadelphia physician and one of the Anti-Federalists' most able economic writers, put it in his *Letter to the Citizens of Pennsylvania*, "we want not that unfeeling plan of Manufacturing Policy which has debilitated the Bodies and debased the Minds of so large a Class of People as the Manufacturers of Europe." No Anti-Federalist, he declared, would welcome the "Manufacturing Capitalist [who] enjoy[s] his luxuries, or fill[s] his Coffers by paring down the Hard-earned Wages of the laborious Artists he employs."[40]

The choices were clear: America could become a nation characterized by unfettered development and rising inequality or one which guaranteed the equitable continuation of independent, small-scale production. Casting their arguments in the rarified light of high political theory, historians have debated the meaning of these choices for more than a decade, concluding that the real choice was one between neo-classical republicanism and liberal individualism.[41] Yet we misunderstand a crucial aspect of these debates if we limit ourselves to the terms of learned political theory. What was at issue in the minds of ordinary farmers and artisans was much more than a contest between traditional republican fears of commercial and industrial development and a liberal creed that championed economic growth of all kinds. At its heart, the dispute was about the continued relevance of their small producer traditions. Philadelphia artisans would take issue with the liberalism of the city's merchants in these years, not because they were traditional republicans, nor because they were liberals of a different persuasion. They would reject the merchant liberalism of the 1790s for the same reasons they had rejected the arguments of forestalling merchants in 1779: because in elevating the rights of non-producers above the rightful claims of the producing community merchants had violated the central tenets of the small-producer creed.

These conflicting visions of American social and economic development first gained national prominence during congressional debates on commercial bankruptcy protection that lasted from 1791 until the passage of the National Bankruptcy Act in 1800.[42] At one level, the issue was whether the federal government should use its power to protect the assets of American merchants during the risky years of the Napoleonic Wars. But at another level, the debate forced the question of American economic development onto the public stage and obliged Federalists and Anti-Federalists alike to define their positions and to choose between mercantile and small-producer images of the future. Speaking for the Federalists, James A. Bayard and Harrison Gray Otis took the side of the merchants, arguing that Americans were in fact "a commercial people . . . precisely similar" to the English. Because of the central importance of commerce in the new nation, they continued, American merchants were a national resource that deserved protection and a chance "to begin the world anew" when commercial disaster struck.[43] The Anti-Federalist spokesman was, again, Pennsylvania's Albert Gallatin, who countered that America, far from being a commercial nation on the British model, was, and would for a long while remain, a small producer's republic where "the same man [is] frequently a farmer and a merchant, and perhaps a manufacturer" in the everyday course of his affairs.[44] If American society remained an undifferentiated order of small producers, the Anti-Federalists reasoned, a commercial bankruptcy law designed to benefit only a small minority of the national community was not only unnecessary but would amount to an unearned privilege given to the nation's merchants.

In the congressional bankruptcy debates as well as in the arguments of Paine, Webster, Otis, and Bayard, Federalists articulated the views of a regnant mercantile class, justified its policies, and set forth a vision of a successful commercial republic that justified their dominion over early national society. For their part, the Anti-Federalists spoke for a more loosely organized rural constituency and envisioned an American future with the family farm at its center. It remained for Philadelphia artisans to fashion their own reply.

III

Although it is often thought of as a process of specious argument and concerted delusion by which a ruling class imposes a set of self-serving views upon its subjects, hegemony is better understood as a form of contentious dialogue spoken between opposing parties. In the early 1790s such a dialogue hardly existed between ruler and ruled in Philadelphia as mercantile power took on a smug, unopposed aspect. True, a few former Constitutionalists continued to rail against the new state and federal constitutions, but until the summer of 1792, they remained isolated and unorganized voices lacking an effective consituency. Counter-hegemony, when it emerged, would have to come from the laboring classes.

The 1780s had been difficult years for the majority of Philadelphia workingmen and their families. The dislocations of the wartime economy, the postwar depression, and the glut of British manufactures in the mid-1780s combined to make the lives of most workingmen increasingly precarious. When overseas demand for Pennsylvania's produce fell dramatically after 1785 and West Indian ports— long the mainstay of Philadelphia's economy—remained closed to American ships, many of the city's craftsmen were driven to the brink of despair. Benjamin Rush recalled that in 1788 fully 1,000 houses lay empty in the city and that "brick layers and house carpenters and all the mechanics and laborers who are dependent upon them" could not find employment anywhere in Philadelphia.[45]

Phineas Bond, the resident British consul and a keen observer of conditions in his native city, noted the impoverished condition of the city's laboring classes as well. In June 1788, he reported on "the wretchedness of the mass of the people here occasioned by the reduced and precarious state of all property."[46] By November conditions had worsened: "I am well convinced," he wrote, that "scarcely an artificer of any sort can at this time meet a decent support." He went on to describe the "great difficulties which many reputable Artificers . . . experience in this country," who "failing in the prospect of employment . . . are left destitute and distressed."[47]

The record of declining wages and soaring almshouse admissions confirm Bond's observations. During the 1780s laborers, mariners, cordwainers, and tailors earned between 20 and 40 percent less for their labor than their prewar counterparts.[48] And as rural and foreign immigrants continued to pour into the economically depressed city, almshouse admissions and the cost of maintaining the poor rose to new heights. By 1787–88 the almshouse population stood at 400, two and one-half times the number seeking aid only four years before.[49] The cost of supporting this destitute population, estimated at 3 percent of the city's total population in 1787, rose accordingly. In 1782–83 the cost of providing food and firewood to the indigent stood at £1,960; by 1786–87 the cost had jumped to £3,893.[50]

In 1788 a broadside addressed "To the Freemen of Pennsylvania" furnished a comprehensive view of the "distresses of the people." Among them were: "amazing numbers of writs and executions [for] trifling sums"; renters "frequently two and three quarters, and sometimes a whole year in arrear"; "tradesmen who formerly . . . were enabled to enjoy the comforts as well as the necessities of life" who cannot now "find as much employment as will produce a bare subsistence for [their] apprentices"; employers unable to pay with punctuality; and laborers who spend in idleness "the season of the year, when most is to be done, and from which [they] have been accustomed to lay up fuel and other necessaries for an approaching winter."[51] The hardness of the times is born out by historian Billy Smith, who found that the lower third of Philadelphia's laboring men and women could maintain themselves in the postwar decade only through periodic use of public and private charity and by resorting to substandard diet and residential overcrowding.[52]

Even when economic recovery finally began in 1790, the city's working poor were slow to feel its effects. In January 1791, the editor of the *Pennsylvania Gazette* could still remark on "the numbers and sufferings of the poor in the City, and especially in the Liberties" where the shipbuilding and maritime accessory trades were concentrated.[53] The return of prosperity seemed even more distant to these workingmen when, in summer 1792, a speculation bubble raised on federal securities collapsed, leaving ships languishing in the harbor and mariners, carters, and all ranks of the support trades destitute and unemployed.[54]

Credit, and with it trade, revived again in the fall of 1792, and with an expanded carrying and re-export trade occasioned by the Napoleonic Wars, Philadelphia's commerce was by 1795 at the highest point in its history.[55] Yet even in the midst of this commercial revival many laboring-class Philadelphians continued to face a harsh world with bleak prospects. The maritime trades may have recovered in unison with the volume of commerce, cordwainers may have benefited from foreign and western demand for their shoes, and the building trades may have benefited from the immigration of the 1790s, but others, among them tailors and common laborers, found relief but little prosperity in the commercial revival of the mid-

1790s. They would hover just above the margins of subsistence well into the nineteenth century.[56]

Faced with these prospects, Philadelphia artisans might well have settled for simple survival, shifting for such temporary employment as could be found, uprooting themselves and their families in search of a competency elsewhere, or perhaps throwing themselves onto the mercies of the city's many private charities. That such roads were taken by some is beyond question; what is remarkable, however, is that so many chose the alternative path of collective organization. It is compelling testimony to the tenacity with which the moral vision of a small-producer community was held by Philadelphia craftsmen that during a long period of adversity they retained and even nurtured a sense of their collective identity and social importance.

This attitude of mutuality and collective self-respect was inscribed within the journeymen's trade and benefit societies that began to appear in the years between the Revolution and the turn of the nineteenth century. Though often overlooked by labor historians, journeymen cordwainers, ship carpenters, printers, stonecutters, masons, carpenters, pilots, mariners, and cabinet- and chair-makers organized on a formal basis during these years, and it is likely that other organizations existed for which no records survive.[57] Patterned after the journeymen's associations of eighteenth-century Britain, these societies provided protection against the misfortunes of everyday life and often served as a forcing ground for early trade unionism as well. Through monthly contributions into a common fund, the societies became a kind of workingmen's bank, providing journeymen with a measure of social security, in the form of sickness benefits, widow's payments, and burial funds, that freed them from the worst insecurities of the age and guaranteed their independence from private charities and the Overseers of the Poor. At the same time, the everyday operations of these journeymen's organizations formalized the personal networks that had long existed among practitioners of the city's trades and helped to produce shop and trade leaders who could be called on when economic downturn or employer assault made collective organization a necessity.[58]

Little has survived to show us the day-to-day functioning of these societies, but price books and charters do provide a small window into the world of the early journeymen's associations. One typical organization was that of the journeymen printers, who formed themselves into a benevolent society shortly after the end of the Revolutionary war. Their early activities are lost to us, but in 1793 they drafted a formal constitution and adopted the name of the Franklin Typographical Society. From their constitution we know that the society provided health, widow's, and burial benefits for its members, and that its treasury served as a loan fund for those who could offer freehold security as collateral.[59] In this, the printers' association was a prototype for the dozens of similar societies that would form in Philadelphia over the coming decades. More revealing, however, was the debate over the society's purpose that occupied its members during the 1780s. Little direct evidence survives, but at some point in the mid-1780s the society divided into two factions, one seeking to limit the association to simple benevolence while the other labored to transform it into a trade union.[60] We know nothing of the heated debates that reportedly took place, but in the end, the trade union faction carried the day, and in June 1786 the Typographical Society called Philadelphia's first recorded strike, standing out against employers who had refused to pay the six-dollar weekly rate that had long been customary in the trade.[61] It was a pattern that would be repeated often in the coming years.

Benevolence and nascent trade unionism do not tell the whole story of the early journeymen's societies, however. In 1795, the journeymen cabinet- and chair-makers responded to the city's wage-cutting outwork system by issuing their own *Book of Prices*. Directed at their employers and the public alike, the journeymen explained the necessity of a mutual list of piece rates because of the "different constructions being put upon [prices] both by employers and journeymen." These conflicting interpretations of just prices had, they complained, "been the cause of frequent disputes between them."[62]

From this brief statement it is clear that one purpose of the journeyman's societies was to negotiate prices and wage rates with masters or merchant employers.[63] This practice had been widespread since colonial times, when it was customary for journeymen and employing masters to contract for wages and piece rates on an annual or semi-annual basis.[64] It was, in fact, a dispute over the terms of their annual contract that had led the journeyman printers to strike in 1786.[65]

Seen in this light, the publication of the *Book of Prices* was less an attempt to increase wages than an assertion of long-held artisan rights. Craftsmen had determined the value of the goods they produced since the Middle Ages, and it was a widely held assumption that the producers themselves knew the difficulty of the work, the quality of the product, and the condition of the trade better than any outsider. The *Book of Prices* was a reassertion of this prerogative in the face of what many journeymen saw as employer encroachment on their natural rights as members of a trade. Coming at the end of the eighteenth century, this defense of customary rights bears witness to the continuing vitality of the small-producer tradition at the very point when employers were beginning to assert their domination over the workings of the craft system. This was a crucial issue in eighteenth- and nineteenth-century England, and equally so among the journeyman cordwainers of Philadelphia who listed employer arrogation of their right to set prices as one of the grievances which led them to form a separate society at the turn of the nineteenth century.[66]

It was the mounting economic pressure of a rapidly expanding capitalist economy, and with it the diverging interests of employing masters and wage-earning journeymen, that led many journeymen's societies to move beyond their benefit and price-setting roles to become trade unions early in the nineteenth century. Again, this process can be followed most closely in the case of the journeymen printers. Informal printers' organizations had a long history, complete with an elaborate ritual—the "chapel"—by which work patterns and social status were decided by members of the trade. In 1802 journeymen printers organized the Philadelphia Typographical Society, and it is their constitution and executive minutes that allow us to enter the world of the city's early trade societies.

The Typographical Society emerged on Washington's Birthday, 1802, in the form of a list of prices directed to Philadelphia's master printers.[67] The purpose of the list, the journeymen declared, was "to be placed on a footing, at least, with mechanics" by having "one uniform price" paid to all journeyman printers in the city. Asserting their traditional rights as members of a very old trade, the journeymen proposed to act equitably, "as men towards men," and expected that their employers' conduct toward them would be "equally candid" in return.[68] The journeymen closed their preamble by emphasizing the mutuality of the trade, hoping that "the time is not far distant, when the *employer* and the *employed*, will vie with each other, the one in *allowing* a competent salary, the other in *deserving* it."[69]

Like the cabinet- and chair-makers before them, the journeyman printers thought themselves no less men and equal citizens for having accepted employment. Their pledge to act "as men towards men" was a reminder that as apprenticed craftsmen they were entitled to the same dignity and respect as their employers. And by reminding their employers of the mutual obligations that existed between master and man, they were reaffirming one of the central tenets of the small producer creed: that skilled and useful labor ought to yield a decent competency and general community respect.

In November 1802, the journeyman printers ratified a constitution and placed their society on a formal footing. According to its charter, the Philadelphia Typographical Society existed "for the mutual benefit and assistance of one another" and "to provide in the day of prosperity for the exigencies of adversity."[70] Under its provisions alimony was to be paid "to sick and distressed members, their widows and children," and money was to be loaned to those "thrown out of employ, by reason of . . . refusing to take less than the established prices."[71]

As its constitution reveals, the Typographical Society was created as a benefit society to provide for the needs of its members in times of distress. But it was also more than this. Along with the city's other journeymen's associations, the Typographical Society was devised not so much to fill a need for social welfare—Quaker Philadelphia was always known for its benevolence—as to escape the dependency that marked any resort to public or private charity. The cornerstone of artisan thought was competency and to be forced to seek relief from the Guardians of the Poor or to apply, hat in hand, for private charity was a direct challenge to one of the artisan's most cherished values. For the artisan, a society in which a man labored hard and honestly and yet could not secure a decent competency was a society severely out of joint. The benefit society represented the skilled craftsman's answer to just such a world and signified a collective attempt to preserve both family independence and the traditional values of the small producer creed.

One final and often neglected feature of the benefit and trade societies was the firmly democratic character of their operations. The Typographical Society was governed by a board of directors, a president, a vice president, a secretary, and a treasurer; and the society's stringent regulations were designed to keep them close to their constituency. The president faced annual election, while the vice president, who was elected by the board of directors, served a term of only four months. The secretaryship was an appointive office while the treasurer, always the most sensitive office in these associations, was selected by the twelve directors from a list of three candidates presented by the general membership. To prevent malfeasance, and the possibility of embezzlement, those directors voting for a treasury candidate became "sureties for the faithful execution of the [treasurer's] duties."[72]

Real authority within the society rested with the twelve directors, who tended to its day-to-day business. Jealous of the power that such a body would hold, the society's constitution provided for an elaborate elective procedure which divided the directors into four classes of three men each, each class to face election quarterly. This complex rotation was designed, the constitution recorded, "so that at every monthly meeting there may be an election for three directors."[73] It was also designed to blur the distinction between leader and led.

What is so critical here is not only the intricate provision for frequent election and rotation in office, but the broad rank-and-file participation that the system implied. It is a glimpse at the workings of everyday democracy, taken from the craftsman's shop and transferred into institutional practice. As such, it was a preview of the political and parliamentary forms that a society of small producers might someday adopt.

IV

In turning to their own collective resources, Philadelphia's journeyman typographers and cabinet- and chair-makers had discovered an important reservoir of strength and determination. In the absence of radical leaders they had created their own institutions, fashioned their own leadership, and taken independent collective action. As other societies began to do the same, it was apparent that the confusion of the 1780s was fast coming to an end.

The laboring classes were not alone in recognizing their growing strength and importance, however. By the early 1790s, the collection of radicals, ex-Constitutionalists, and disaffected Federalists who were beginning to form a national opposition party made their first attempts at alliance with Philadelphia's growing journeymen's movement. As they soon discovered, healing the wounds inflicted at Fort Wilson would not be an easy task. The memory of Fort Wilson remained vital in the 1790s and the resentment felt by Philadelphia craftsmen toward apostate radicals continued to smoulder. Even more critical, the political record of the 1780s had been one of near-total neglect and did little to smooth the path toward reconciliation. In the 1790s, a successful alliance would require substance, not party rhetoric.

Opposition to Federalism from the ratification of the Constitution in 1788 until the revival of popular politics in 1792 was a private and personal affair. Shunned by city artisans after their ill-fated attempt to pass an effective tariff in the mid-1780s and defeated in their efforts to halt the conservative restoration at either the state or local levels, Philadelphia's Constitutionalist party dissolved, leaving its leaders isolated and powerless. The revival of political opposition, which came slowly to the city in the summer of 1792, was largely the fruit of one man's labor.[74] From his return to Philadelphia in 1777 to become Pennsylvania's surgeon general, Dr. James Hutchinson had been a committed Whig and a dogged political worker among the ranks of Philadelphia craftsmen.[75] In the constitutional debates of 1787 and 1788 he had been the bulwark of anti-Federalism in Philadelphia, writing the "OLD WHIG" essays that helped to focus mounting city opposition to the federal Constitution.[76] With the Federalist triumph in 1789, Hutchinson became the natural leader of opposition in the city, corresponding with radicals throughout the state and agitating among the Irish and Scots-Irish immigrants who flowed into the city after 1783.[77]

Under Hutchinson's guidance, this postwar Irish immigration played a crucial role in forging a new radical–laboring-class alliance in Philadelphia. Precise immigration figures do not exist before 1819, but surviving records suggest that nearly 26,000 Irish immigrants settled in Philadelphia between the end of the Revolutionary war and the turn of the nineteenth century.[78] Unlike earlier Irish immigrants, who had come as indentured servants, most postwar immigrants were established farmers and artisans who paid their own passage to America.[79] And while farmers made up a majority of the more than 53,000 Irish immigrants who arrived in Philadelphia between 1789 and 1806, a large minority—roughly 30 to 40 percent—was composed of Ireland's "best artisans with their families."[80] Counted among these more prosperous craftsmen were also those described by an Irish customs officer as "the lower order of tradesmen," the tailors, smiths, joiners, and others who made up the city's pedestrian trades, as well as substantial numbers of weavers who would become an important element in Philadelphia's expanding textile industry.[81] Together, these Irish craftsmen and their families flooded into Philadelphia's suburbs, and especially the southernmost Southwark district, turning it from a moderate-sized shipbuilding satellite into a thriving Irish laboring-class quarter.[82]

In addition to recomposing Philadelphia's laboring class, Irish immigration also provided an important constituency for the city's infant opposition movement. Many of these immigrant craftsmen came to Philadelphia with the twin experiences of English repression and the failed United Irish independence movement fresh in their minds. One Irish customs official noted the radicalism of these departing artisans when he reported that "many of [the workmen] who leave are of discontented and turbulent dispositions"; the very kind of people he thought were best left to emigrate to American shores.[83] As important as the indigenous radicalism of the Irish craftsmen was the fact that a large number of United Irish leaders came to Philadelphia as well, making it, in the words of one of their number, "the most respectable emigration which had taken place to the United States since the settlement of the New England Colonies."[84] Together these Irish craftsmen and political leaders injected a new and more militant form of radicalism into the mainstream of Philadelphia politics. Combining a revulsion toward anything British with a profound commitment to political struggle against economic exploitation and aristocracy, these Irish craftsmen and middle-class radicals formed a well-prepared ground from which to raise a movement directed against the city's Federalist elite. It was thus no accident that Irish exiles such as John Binns, Mathew Carey, and Irish-reared William Duane would play such an important role in Philadelphia politics from the 1790s until the end of the 1820s.[85]

While James Hutchinson stirred the fires of opposition among immigrant workingmen, three other men emerged who, together with Hutchinson, formed the core of early anti-Federalism in Philadelphia. The most prominent of these three was the Jamaica-born attorney Alexander James Dallas, who arrived in Philadelphia in 1783. When the postwar depression kept his early practice modest in size, Dallas spent much of his time editing the *Pennsylvania Evening Herald* and climbing the rungs of Pennsylvania's political ladder. In 1790, Governor Thomas Mifflin acknowledged Dallas's political ascent by appointing him Secretary of the Commonwealth, a position which he used to promote his own political fortune as well as the interests of the fledgling opposition party.[86]

If Hutchinson and Dallas were the driving force of anti-Federalism in Philadelphia during the early 1790s, Jonathan Dickinson Sergeant and Blair McClenachan provided vital support. Having served as New Jersey's delegate in both Continental Congresses, Sergeant fled to Philadelphia when Hessian troops burned his Princeton home in 1776. Once in the Quaker City, he joined the ranks of the Constitutionalists and served as attorney general of

Pennsylvania from 1777 until 1780. Like Dallas, a member of the Philadelphia bar, Sergeant added both his political experience and his prestige as a Revolutionary patriot to the opposition cause.[87] Blair McClenachan was an Ulster-born merchant of decidedly democratic sympathies. Active in the patriot cause throughout the Revolutionary era, McClenachan became an early Anti-Federalist and, in the first federal elections of 1789, ran for Congress as the city's opposition candidate.[88] Despite his ubiquitous presence in democratic politics, however, relatively little is known of his actual activities. Apparently he spent much of his time organizing among fellow merchants and was instrumental in bringing dissident merchants into the Democratic-Republican party during the mid-1790s. By reputation a fiery orator, McClenachan was also well known for organizing and leading Philadelphia's most spectacular anti–Jay Treaty demonstration.[89] It was the labor of these four men— Hutchinson, Dallas, Sergeant, and McClenachan—that accounted for the rise of the Democratic-Republican party in Philadelphia and, in the process, established a new political vehicle for the expression of laboring-class concerns.

The first serious attempt to reunite middle-class democrats and Philadelphia artisans came during the congressional campaign of 1792.[90] In July of that year Philadelphia Federalists had put into motion a plan designed to assure their control of the state's nominating process. The plan called for the creation of local Committees of Correspondence in which the "best" men in each county would meet to select delegates to a statewide nominating convention. The convention would, in turn, draw up a list of candidates to be presented to voters in the fall elections. Expecting little opposition from a divided and politically moribund popular movement, the Federalists sought to push their proposal through a lightly attended town meeting on the night of July 25.

But if the Federalists thought they could attach a popular imprimatur to their plan so easily, they were profoundly mistaken. When James Wilson, chairman of the July 25 meeting, read the names of Philadelphia delegates from an obviously prepared list, the expectedly quiescent town meeting broke into an uproar. First one, then several voices from the crowd demanded a discussion of the candidates, and soon many began questioning the conduct of the meeting itself. In the end, Wilson was forced to adjourn and set a new meeting for July 27.

Hutchinson and Dallas, who had attended the July 25 meeting, saw it in its true light: they noted that the proffered delegates were mostly the same men who had endorsed Federalist Arthur St. Clair for governor a few months earlier. Suspecting the worst, Hutchinson and Dallas worked feverishly "to rouse the people to support their independence, and to think & act for themselves."[91] The working-men of the city took this advice to heart. At the July 27 meeting they voted down the Federalist plan by a two-to-one majority.

This was not the end of the matter, however. Taken aback by their defeat and the sudden revival of popular assertiveness, the Federalist chairman, state senate speaker and former mayor Samuel Powel, unexpectedly declared the sense of the meeting *in favor* of the Federalist plan and hurriedly adjourned amidst hisses and shouts from the crowd. In response to this violation of popular sovereignty and simple common sense, Hutchinson issued a call for another town meeting to be held on July 30.[92]

Considering the quiescence of the post-ratification political scene, it was probably to everyone's surprise when 2000 workingmen turned up in the State House Yard. Describing the scene to Albert Gallatin, the opposition's western leader, Hutchinson reckoned that the meeting was the largest public gathering in Philadelphia since the price-control controversy of 1779.[93] In attempting to circumvent the democratic process and force a pre-selected slate of candidates on the city's voters, the Federalists had betrayed their own devious purposes and opened the gates of reconciliation to Philadelphia radicals and workingmen. When, on July 31, the Federalists again attempted to circumvent the popular will by calling yet another town meeting—this one at three o'clock in the afternoon, when most mechanics were still at work—and placed Robert Morris in the chair without a popular vote, a group of workingmen stormed the podium, seized the speaker's chair and table and in a gesture that foretold the future of Philadelphia politics, demolished them to the approving cheers of the crowd. As Hutchinson wrote of the incident a few weeks later, "it was with difficulty violences of a more serious nature were prevented."[94]

Following the July 31 meeting, the Federalists returned to safer ground and pursued their nomination scheme in small meetings of their already committed supporters. The planned nominating convention was eventually held on September 20, 1792, but representatives from only half of the state's counties took the time to attend. The Federalist plan was a failure. For their part, Hutchinson and the city radicals were now free to put their own plan into effect. Approved at the July 30 public meeting, the radical plan called for a general poll of the state's voters to be conducted through a series of local town meetings. Popular response to the plan was gratifying, and the radicals demonstrated their own democratic principles when, in late September, they published the results of the canvass, even though it contained the names of both Federalist and anti-Federalist supporters.[95]

The growing opposition movement did well in the election of 1792. In addition to electing three of their western leaders to Congress, the movement elected John Swanwick, a Philadelphia "merchant-Republican" recently won over from the Federalists, to the state assembly.[96] Reviewing the election for Gallatin, Hutchinson wrote that "notwithstanding however the strong opposition, we had a majority in Philadelphia County for the whole, and were close on the heels of our Opponents in the City."[97]

The result of the 1792 election was a clear sign that the city's laboring classes were, however tentatively, again accepting the political leadership of men above them in station. The strong support that the Democratic-Republicans received from the laboring-class suburbs suggests as well that Philadelphia workingmen saw in the emerging Democratic-Republican party a truly popular opposition movement. Hutchinson and Dallas had labored hard to present just such an image to the city's workingmen, naming their ticket after France's revolutionary "Rights of Man" and openly attacking Federalist actions as a "consummate display of the presumption and boldness of the Aristocratic junto."[98] Finally, addressing the "Freemen of

Pennsylvania," the Democratic-Republicans spoke directly to small producers throughout the state:

> The times are changed, it is true, but you are not changed with them. You see elevated into great official distinction, many, who did not dare to appear in public view, at the period which "tried men's souls," and an attempt is made to dazzle you with that blaze of wealth, which the fruitful agencies of the war, and the rank speculations of the Fund, have made more mysteriously than honestly accumulated in the hands of individuals. But you are not by such circumstances to be awed into silence, or deluded into slavery. The spirit of the Revolution is not dead, though perhaps it sleepeth. Those minds which the splendor, the power, the influence, and the corruption of the British Monarchy could not captivate, subdue, or contaminate, must be superior to the arrogance of an upstart Aristocracy, the machinations of the inveterate foes to independence, and the soothings of those crafty politicians, who aim at a monopoly of power.   "MENTOR"[99]

"Mentor"'s aim was exact; in capturing the essence of popular feeling against the turn of the times, he articulated publicly what many had privately felt. The impressive laboring-class attendance at the July meetings was but one indication that Hutchinson and Dallas had struck their mark.

If the battles of 1792 had set the stage for the emergence of a revitalized popular movement, it was the French Revolution that smoothed the way for reconciliation between radicals and the city's working classes. News of the fall of the Bastille did not appear in the Philadelphia press until late September 1789, and the first reactions in the city were overwhelmingly ones of support for what was widely believed to be a movement for liberal reform. But by late 1792, as the truly revolutionary nature of the events in France had become apparent, pro- and anti-revolutionary sides began to be drawn in Philadelphia. The first popular pro-French demonstration took place on New Year's Day, 1793, and from that time onward a Philadelphian's political character—whether he was judged an aristocrat or a democrat—was measured by the strength of his support for the French Revolution.[100]

Enthusiasm for the Revolution was especially strong among the laboring classes of the city, who embraced it as the child of their own Revolutionary hopes. Philadelphia artisans sang the "Marseillaise" as they passed through the streets of the Federal City and hung on every word printed about events in revolutionary France. The French cockade quickly became a staple of laboring-class garb and for months after the king's execution, hundreds of working people lined up at the Black Bear tavern in Second and Callowhill streets to witness a graphic re-enactment of Louis XVI's demise. When, at the end of each performance, the king's head rolled into a straw basket and his lips turned a final shade of blue, the Black Bear shook to the roar of stomping feet and rousing cheers from the laboring-class crowd.[101] In these and many other ways, laboring-class Philadelphians took the French Revolution into their hearts and made it their own.

This enthusiasm was shared broadly, if less demonstrably and in a more refined voice, by middling- and upper-class anti-Federalists who saw in the popular radicalism revived by the French Revolution both an opportunity to strengthen their oppositional alliance and a trans-Atlantic confirmation of their anti-aristocratic principles. Overwhelmed by the anti-authoritarian atmosphere, city Federalists found themselves isolated from the tide of popular events and could only grumble and rail against the times. Their attitude was captured in a toast offered at Richardet's tavern, the local meeting ground for Federalist merchants and politicians, on the birthday of George III: "To the Cap of Liberty, may those who wear it know that there is another for licentiousness."[102]

It was opportunity and not licentiousness that opposition leaders saw in Philadelphia's revived popular movement, and to tap the power of popular anti-aristocratic feeling they created the Democratic Society of Pennsylvania. Formed in May 1793, the Democratic Society was the creation of men from the middling ranks of Philadelphia's merchant and professional community.[103] Led by David Rittenhouse, Charles Biddle, Peter DuPonceau, Israel Israel, Benjamin Franklin Bache, and Michael Leib, the Society brought together men with decidedly pro-French and strongly anti-Federalist sympathies.[104] Nominally independent of the fledgling Democratic-

Republican party, the Democratic society nonetheless counted Hutchinson, Dallas, Sergeant, and McClenachan among its earliest members. Minutes for the first few months of the Society's existence have not survived, but by the winter of 1793 they document a growing radical wing within the Society which, led by Hutchinson, sought to mold the Society into an effective arm of the emerging opposition party.

What drew many Philadelphia artisans to the Democratic Society was more than political interest, however. It was shared membership in Elhanan Winchester's staunchly anti-Calvinist First Universalist Church. The religious connection was an important one. In an era of widespread spiritual experimentation that witnessed the rise of Methodist, Baptist, and Presbyterian evangelicalism as an alternative to traditional artisan values and saw Benjamin Franklin Bache distribute 15,000 copies of Paine's deistic tract *The Age of Reason* from his shop on Market Street, Universalism offered a religion that was remarkably in tune with the small-producer traditions and the diffuse religiosity of Philadelphia artisans.[105] Rational, yet solidly Christian in doctrine, Universalism rejected Calvinist notions of election and eternal damnation as cruel and authoritarian at the same time that it accepted the Bible and denied the extreme anti-Christian aspects of radical deism.[106] Most important, Universalism literally resonated with traditional artisan values. In Winchester's popular hymn "America's Future Glory and Happiness," for example, the Universalists focused on competency and economic independence:

No more the labour'r pines, and grieves,
For want of plenty round;
His eyes behold the fruitful sheaves,
Which make his joys abound.[107]

By echoing the traditions of the craft community, Universalist hymnody captured the sentiments of Philadelphia artisans and gave a new voice to their most deeply held values:

How sweet is the union of souls,
In harmony, friendship and love;
Lord help us, this union to keep,
In *union* God grant we may meet.[108]

Written at a time when city journeymen were hotly debating the value of trade-union organization, the idea of union expressed here powerfully linked together the separate communities of craft, church, and early trade union.

Winchester's Lombard Street church brought together artisans, radicals, and a gathering of liberal intellectuals who came not only to hear Universalism's message of democratic salvation but to discuss ways in which they might apply their moral precepts to the political world around them. Nearly 40 percent of Winchester's church members were active members of the Democratic Society, and this shared membership helped to close the gap that had existed between artisans and opposition leaders since the defeat at Fort Wilson fourteen years before.[109]

The *Principles of* the Democratic Society, published in June 1793, reveal its ultimate political purpose. Addressed to "every virtuous citizen" of Pennsylvania, the *Principles* attacked the arrogance and self-interest of Philadelphia's Federalist ruling class.[110] In contrast to the orotund rhetoric of high Federalist pronouncement, the tone of the *Principles* was that of Thomas Paine. It began by recalling the achievements of the American and French revolutions, which it claimed had taught ordinary people to be neither "dazzled with adventitious splendor" nor "awed by antiquated usurpation."[111] After reiterating the old Leveller notion that the value of freedom and equality was "best understood by those who have paid the price of acquiring them," the *Principles* went on to caution its readers that the achievements of one generation have "too often been lost by the ignorance and supineness of another."[112] It was the declared purpose of the Democratic Society to prevent such a declension by surveilling "public Servants" and by aiding the people in selecting men "according to their intrinsic merits . . . regardless of party spirit or political connection."[113]

These are unarguably the principles of a political society, a protean opposition party in all but name. In its rhetoric can be found the rationalism and radical republicanism of England's Society for Consitutional Information and its Philadelphia offspring, the Society for Political Enquiries, an organization founded in 1786 to study the science of politics. These were all societies of opposition, surveillance, and reform that defended notions of popular sovereignty and civil rights in the dangerous final decade of the eighteenth century. But by their appearance in Philadelphia during a period of active reaction and popular agitation, the principles of the Democratic Society were also an appeal to city workingmen. The "adventitious splendor" and "antiquated usurpation" condemned in the society's preamble may have referred directly to the mechanisms of monarchy and deference in Europe, but they were also a more subtle reference to the display of wealth and aristocratic pretension among Philadelphia's ruling class. The parenthetic aside which claimed that the most important social values are "best understood by those, who have paid the price of acquiring them" is an unmistakable gesture toward the laboring classes who populated the Revolutionary militia and who had watched their depreciated certificates of compensation make effortless fortunes for already rich men.

But it was in the fifth, and final, of its principles that the leaders of the Democratic Society made their most overt appeal to the laboring classes of the city. In words that echoed the values of the small-producer creed, the people of Pennsylvania were said to "form but one indivisible community whose political rights, interests . . . and prosperity must in degree and duration be forever the same."[114] Much more than a specious rhetorical flourish, the drafters of the *Principles* were here paying homage to both the continuing strength of small-producer traditions and the growing power of the city's laboring-class movement.

Yet though many of its *Principles* were addressed to Pennsylvania farmers and mechanics, in practice, the Democratic Society was far from being a small-producer organization. The Society's membership lists reveal a preponderance of merchants, professionals, and higher-ranked artisans joined together with a handful of government officials and merchant's clerks.[115] Only 74 of the 217 members listed themselves as craftsmen. And while this amounted to roughly one-third of the total membership, it is unclear how many of these craftsmen were employing masters rather than dependent journeymen. More than one-third of the laboring-class members did come from the lower-ranked occupations—mariners, laborers, cordwainers, weavers, and tailors—although some of the cordwainers, weavers, and tailors were undoubtedly masters. Thus, more than anything else, the Democratic Society resembled the social composition of the popular committees of the Revolutionary era.

The fact that the Democratic Society did not have a large laboring-class membership is less important, however, than its existence as a vehicle which maintained contact between Philadelphia's opposition leaders and the city's working classes. These personal and ideological links were further reinforced in the fall of 1793 when demographic catastrophe struck the city. Since 1791 slave rebellions on the West Indian island of Saint-Dominque had driven hundreds of slave-owning French planters into Philadelphia seeking refuge. In addition to their luggage and household effects, these West Indian refugees arrived in the city carrying an even greater burden—the affliction of yellow fever.

The history of the epidemic of 1793 is a famous one and need not be recounted here.[116] The epidemic was truly of catastrophic proportions, and its horrors can be followed in Arthur Brockden Brown's *Arthur Mervyn* as well as in Mathew Carey's more statistical account. Of greater importance than the record of human suffering—which in Philadelphia occurred on an unprecedented scale—was the impact of the epidemic on the city's laboring classes and their connection with the emerging Democratic-Republican movement.

Almost as soon as the fever was diagnosed by Benjamin Rush in early September, the city's elite began to flee. While a few merchants and public officials remained, by mid-September Philadelphia was a city of the dead and dying—and the laboring classes. While families of moderate or more substantial station could retreat to country estates or board out with rural relatives, few laboring-class families had such options available to them. Their fate lay in the streets and alleyways of Philadelphia.

Workingmen and their families paid dearly for their sojourn in the city that fall. When Mathew Carey tallied the tables of mortality he had so meticulously compiled, he found that nearly 85 percent of those who died came from the ranks of the mechanics, journeymen, and laborers who had remained in the city.[117] Most prominent among them were the recent Irish immigrants who formed the backbone of Democratic-Republican support.

The impact of this catastrophe on the minds of Philadelphia working people can only be imagined. Certainly no family lived through the two wretched months of September and October untouched by illness, death, or wrenching psychological trauma. For many like John Cox, a city shoemaker and amateur poet, it seemed that the world had come to an end:

To see the streets at noon—a most solemn sight—
They looked as dreary as dark midnight;
The cries of dying souls drove to despair,
With frightful shrieks they rent the very air.
Malignant fever stalk'd with haughty pace,
And thousands yielded to death's cold embrace.[118]

Yet despite the unparalleled misery of these months, the working people of Philadelphia were not blind. They had witnessed the Federalist elite, who claimed to represent the community interest, flee the city with hardly a thought to the plight of those who were forced to stay behind. By the second week in September, so many of the city's elite had moved outside Philadelphia that Mayor Matthew Clarkson could not even muster enough public officials to run the municipal government.

The wealthy Federalists' desertion of the city and their callous disregard for the well-being of the working community served only to reinforce what Philadelphia radicals and working people had thought all along: the concerns of the city's men of wealth extended no farther than their own interests and the collective well-being of their class.

If the nomination battles of 1792 and support for the French Revolution had brought Philadelphia's radical leaders into contact with a potential laboring-class constituency, Federalist mean-spiritedness and their callous disregard for public safety during the yellow-fever epidemic provided the adhesive sentiments that would bind them together. Although the motor of city opposition in the early years of Federalist dominance—Dr. James Hutchinson—died of the fever in 1793, new leaders such as Dr. Michael Leib and Israel Israel quickly rose to take his place. The vehicle that created this new leadership was, ironically, the vacuum left by the Federalist exodus. [119]

With the city government in self-exile, Mayor Clarkson turned to the people themselves for the governance of the city. Organizing an ad hoc, extra-legal Committee (it had no, more formal name), Clarkson turned the city into a commune. The Committee, which operated from September 14 until the end of the epidemic on October 31, was composed of merchants who made up its officers, professionals and master craftsmen who headed the working committees, and ordinary mechanics who performed the delegated work. [120] The Committee, like the Democratic Society to which a quarter of its members and all but one of the correspondence committee belonged, was a replica of Philadelphia's Revolutionary government. [121]

The Committee took up the everyday tasks of city government, but, under the guidance of Israel, Leib, and the French-born merchant Stephen Girard, especially concerned itself with caring for the sick, providing for the growing number of orphans, and burying the dead. Like the popular-radical Committee of Trade in 1779, it also regulated the price of gruesome necessities such as nursing, coffin-making, and public burials. For laboring-class Philadelphians, especially those in the lowest ranks, the Committee's work was the only sign of public concern to be found during those terrible months. When the city revived later in the year, these same mechanics and laborers would remember with genuine thankfulness the peripatetic ministerings of men like Stephen Girard, Israel Israel, and Michael Leib; and they would remember with matching rancor the callousness of Philadelphia's self-exiled elite.

In January 1794, as the city began to recover from the fall epidemic, the Democratic Society moved even closer to the city's laboring-class voters by electing Blair McClenachan to its presidency. McClenachan was a popular figure in laboring-class circles and his election as head of the Democratic Society was a clear attempt to incorporate workingmen into Democratic-Republican politics.

McClenachan began to forge links between the Democratic Society and city craftsmen shortly after his election. On May 8, 1794, he joined over 300 journeymen in the State House Yard to protest recent excise taxes introduced by the Federalist Congress. [122] Then, at the end of the month, as the Jay Commission departed for England, McClenachan helped organize a street demonstration in which an effigy of Jay was "ushered forth from a barber's shop amidst shouts of the people," paraded through the city, guillotined, and blown up with gunpowder. [123] Later in 1794 and again in 1795, he moved the workingmen of the Democratic Society out into the streets to protest other Federalist policies.

A new alliance did indeed appear to be forming in the mid-1790s, centered on support for France's popular revolution and hatred of Britain and all forms of aristocratic government. The Bastille Day celebration of 1794 underlined this convergence as Philadelphia radicals and workingmen joined together for an evening's drink and entertainment that was supplied not by the radicals, but by the city's shipwrights and carpenters. [124]

But while contacts were made, friendships cemented, and the points of alliance ironed out between the city's radicals and workingmen between 1794 and early 1795, it was news of Jay's pro-English treaty that brought the two sides closer together than they had been since the early years of the Revolution. Senator Stevens Thompson Mason, a Virginia Democratic-Republican, leaked the terms of the treaty in June 1795, after which it became the subject of heated controversy and opposition throughout the young nation. The impact of Jay's Treaty was especially profound in Philadelphia where the *Independent Gazetteer* described the year's Independence Day celebration in the darkest terms: "the birthday of American liberty was celebrated in this City with a funeral solemnity," the editor reported; "it appeared more like the internment of freedom than the anniversary of its birth. The countenances of the citizens generally appeared dejected, and the joy and festivity which usually characterize the day seemed to be superseded by sadness."[125]

The depth of popular feeling against the Jay Treaty, and the capitulation to aristocratic government that it signified, can be seen most powerfully in the crowd action that closed the Independence Day celebrations that year. Toward evening a crowd of "citizens" gathered in the laboring-class district of Kensington, hoisted a transparency of Jay onto their shoulders, and began to march toward the city. The transparency expressed their feelings graphically and unmistakably. Jay was represented in his judicial robes holding a scale, one side "containing *American Liberty and Independence*, kicking the beam," the other holding "*British gold* in extreme preponderance."[126] The visual message was underlined by a projection emanating from Jay's mouth which read "*Come up to my Price and I will sell you my Country.*"[127]

The procession moved peacefully through the city after which it turned and marched back to Kensington; there the crowd burned Jay's effigy in a bonfire that could be seen from Southwark, at the southern edge of the city. By midnight, Federalist magistrates had seen enough and called out Captain Dunlap's Light Horse to put an end to the demonstration.[128] When too few of its elite members could be found, the task fell to Captain Morrell's cavalry detachment. Faced with an "invasion" by Morrell's Federalist troops, the Kensington crowd chose to stand ground and fight. In the ensuing battle Morrell's cavalry was overpowered and forced to retreat into the city. At that point, city officials apparently decided that, lacking sufficient force to disperse it, the crowd would be left to continue its revels. To commemorate their victory, some men from the crowd erected a signpost on the site of the encounter; it read: "Morrell's Defeat—Jay Burned—July 4, 1795."[129]

The events of July 4 were emblematic of popular dissatisfaction not only with the Jay Treaty but with aristocratic rule in general. In their demonstrations the people of Philadelphia did not seek to make fine distinctions between the aristocratic government of England and that of their own city, state, and nation. The intention of Philadelphia Federalists to tolerate neither plebeian insubordination nor popular criticism of their policies was manifest in the marshaling of Morrell's Light Horse against a peaceful and orderly crowd. The same arrogance and concern for power would surface again in 1798, when President John Adams celebrated Bastille Day by signing the Alien and Sedition acts into law.

Extreme events such as those of 1794 and 1795 are often our only access to popular thoughts, notions, and visions. In these years that witnessed the birth of organized opposition to the policies and pretensions of Federalist merchants and their allies, we also find a wider reservoir of opposition in the hearts and minds of Philadelphia craftsmen. The extent to which the two sources of opposition joined in defiance of Pennsylvania's restored ruling class and its increasingly anti-democratic actions cannot be measured with any degree of accuracy. But it is certain that a dialogue was begun in that direction. The reply of the democratic celebrants of Independence Day, 1796, when they were asked by Governor Thomas Mifflin and the Society of the Cincinnati to hold off their cannon salutes until that august company had finished their round of toasts, reflected the stubborn-minded values of city workingmen. Faced with men of overbearing wealth, power, and social position, their answer was as forceful as it was direct: "as they were called upon to honor the day," they told Mifflin, "they as freemen and soldiers, were not bound to wait on any description of men."[130]

Artisan workshops were governed by norms of mutuality, cooperation, and common purpose even in the largest early American enterprises, such as the shipyard illustrated here.
(*Courtesy, Historical Society of Pennsylvania*)

BLACKSMITH

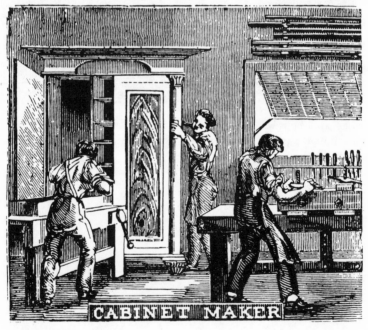

CABINET MAKER

(*above*) Many artisans relied on their families as an ordinary part of production. Here in a typical weaver's home, a mother and daughter spin the yarn that the husband will weave into finished cloth.

(*above, left*) Cooperative labor was both a necessity and a norm in the pre-machine age, as illustrated in this view of blacksmiths at work.

(*below, left*) The division of labor was a characteristic of many artisan workshops long before the advent of the manufacturing system. Here skilled cabinetmakers work independently at individual tasks before joining the separate parts together to create the finished cabinet in the background.

"Zion Besieged and Attacked," 1787. The radically democratic Pennsylvania Constitution of 1776 came under increasing attack during the post-Revolutionary years. In this early political cartoon, artisans play a prominent role in defending the state constitution against merchants, bankers, and the state's traditional elite. Here the artisans (mechanics) occupy the left ramparts of the fortress at the upper right, while the conservative elite (led by Robert Morris on horseback at left center) attempt to storm the constitutional citadel.
(*Courtesy, Library Company of Philadelphia*)

William Duane (1760–1835), editor of the radically democratic newspaper the *Aurora* and author of the influential "Politics for Farmers
and Mechanics" essays, was the foremost spokesman for Philadelphia
artisans before the 1820s. He is captured here in 1802 by the French
portraitist Fevret de St.-Memin.
(*Courtesy, National Portrait Gallery, Washington, D.C.*)

"The Ghost of a Dollar or the Banker's Surprize," 1806. Banks were viewed with great circumspection by artisans, who saw them as bastions of elite power. Artisans also suffered from the lending policies of most banks, which discriminated against those with small capital. The banker lampooned here as "Stephen Graspall" was Stephen Girard, Philadelphia's most prominent banker and the city's wealthiest man. (*Courtesy, Library Company of Philadelphia*)

"Procession of Victuallers," 1815. Although the solidarity expressed here by the ordered ranks of craftsmen was already being eroded by divisions between masters and journeymen, processions such as this one held at the close of the second Anglo-American war offered artisans an opportunity to display their craft pride and to remind the public of the importance of their labor.
(*Courtesy, Historical Society of Pennsylvania*)

"Democracy Against Unnatural Union," 1817. The end of the second Anglo-American war brought renewed political infighting between the merchant-manufacturer and laboring-class wings of Philadelphia's Democratic party. In this cartoon, the pro-capitalist New School lampoons Michael Leib's attempt to retain Old School power by allying with local Federalists. Leib, who had been organizing city artisans since the 1790s, is shown at the right, supported by venial office-seekers and Duane's *Aurora*.

(*Courtesy, Library Company of Philadelphia*)

Chapter 5

An Apprenticeship to Class, 1796–1810

The period between 1720 and 1795 marked the youth of Philadelphia's laboring-class movement. It had been a period of growth, learning, and collective definition. Growth had come from the high birth rate and healthy longevity of its native-born residents augmented by the flood of English, German, and Irish immigrants who flocked to the city along with rural migrants from the Pennsylvania countryside. Learning had come from everyday experience and from the tutorage of patricians and people of middling circumstance who sought to extend the realm of political participation to include the ranks of the laboring classes. More important, participation in the militia and membership on the radical committees of the Revolutionary era brought a new sense of self-respect to the working people of Philadelphia and taught them lessons about their real power within the community. To these changing ideas and experiences, workingmen brought the heritage of their craft culture, bound together by a small-producer tradition that emphasized useful labor, independence, equality, and competency and provided them with an independent standard of moral judgment.

With the growth of independent journeymen's organizations after the Revolution and the development of recognizable laboring-class districts along the fringes of the rapidly expanding city, Philadelphia craftsmen entered a period of apprenticeship in which they would hone their intellectual, political, and organizational skills, and in the process transform their small producer heritage into a culture that was becoming recognizably working-class.

I

The events of the early 1790s had demonstrated the internal strength and political potential of Philadelphia's craft community, and the city's Democratic-Republican leaders lost little time in bringing workingmen into the fold. After securing a working alliance with disenchanted merchants, manufacturers, and liberal professionals, city Democratic-Republicans began, in 1796, to turn their attention increasingly toward the laboring classes. The first sign of this strategy came at the beginning of the year when John Beckley, an early Democratic-Republican leader and future manager of Jefferson's 1800 campaign, added "stock-jobbers [and] speculators," a traditional animus of city craftsmen, to the list of Democratic-Republican enemies.[1] A week later "Libertas" added his own injunction that linked small-producer and Democratic-Republican principles: "The Laborer is worthy of his hire—reward him with Liberality. The King and the sycophant are useless reptiles—continue to despise them."[2]

In their appeals to small-producer principles, Beckley and "Libertas" were preparing for the 1796 election, which would be the first real test of Democratic-Republican strength in Philadelphia. If the Democrats could add laboring-class votes to those of opposition merchants, manufacturers, and professionals, they would be in a decisive position to overturn Federalism in the Quaker City. Hope ran high in Democratic-Republican circles as local politicos heralded Independence Day with a toast to "The election of 1796—May all the officers of Government be cast in a pure Democratic mold."[3] The real work began in September, the traditional time for compiling election slates. Meeting late in the month, city Democratic-Republicans extended their laboring-class appeal by placing Anthony Cuthbert, Thomas Leiper, Jacob Bright, and Matthias Sadler, all master employers, on their slate of common council candidates, carefully identifying each candidate by his trade.[4] Likewise, in the critical laboring-class suburb of the Northern Liberties, the Democratic-Republicans took great pains to nominate candidates with broad laboring-class appeal, choosing Emanuel Eyre, a local shipbuilder, Blair McClenachan and Michael Leib, two

Democratic-Republican leaders with intimate ties to district workingmen, and George Logan, a popular spokesman against Federalist trade and economic policies, to represent the district in Congress and the state assembly.[5] In early October "An Elector" sought to insure laboring-class identification with the Democratic-Republican cause by putting forward a revised slate of candidates which included Israel Israel's name along with a pointed reminder that many on the slate had "great and peculiar claims upon the generosity and gratitude of the citizens of Philadelphia for their exertions during the Yellow Fever of 1793."[6] As "An Elector" well knew, of the "citizens" he mentioned, most were members of the city's laboring classes.[7]

While Philadelphia Democratic-Republicans busied themselves with public and private appeals to laboring-class voters, "A Mechanic" addressed the concerns of his fellow workingmen directly. Although anxiety about the aristocratic designs of the Federalists at the state and national levels was good and proper, he argued, the operation of the city's aristocratic Corporation had a more immediate claim on the workingman's attention. To begin with, he noted that the Corporation had not published a public accounting of its finances, something it was by law and custom required to do. Workingmen were taxed by the Corporation, he declared, but were not considered important enough to be informed about the uses to which their tax money was put. This led "A Mechanic" to speculate about the Corporation's motives. To understand why city finances were such well-kept secrets, he suggested, one had only to examine the situation of the duties charged for wood landed at the public dock. The bulk of these collections, he informed his readers, did not find their way into the public treasury at all, but went instead into the pockets of "those to whom the collection was farmed out" by the members of the Corporation.[8]

More serious, and certainly more immediate, than the Corporation's questionable financial practices, however, were a number of regulatory ordinances which they had passed during the early 1790s. These ordinances, ranging from restrictions on the display of goods to the building of wood-frame houses, intentionally struck at traditional craft practices and at the laboring-class way of life.[9] The ordinances were blatantly hegemonic, designed to impose a ruling-class view of order and cleanliness upon the city and, some workingmen claimed, were ultimately contrived to drive workingmen and their families from the city entirely. But severe as they were, the ordinances were mild compared with the officers appointed to enforce them. "A Mechanic" noted that "the ordinances respecting high constables have been much complained of" by city workingmen who "have always considered them as worse than useless; much the greatest part of their emoluments have arisen from heavy, and in some cases illegal, fines imposed by the Corporation for very trivial faults."[10] Indeed, according to "A Mechanic," many workingmen thought that the constables had been appointed "for no other purpose than to act as spies upon the Citizens [for crimes] too trivial for the notice of a liberal police."[11]

Much the same could be said for the numerous "ordinances for the suppression of nuisances and enforcing useful regulations" which extended the reach of the Corporation into the minutiae of everyday life. "A Mechanic" lamented that "these fines (found illegal by the Supreme Court) were enforced by the Aldermen with so much severity, and collected by the high constable with so much avidity, that it was supposed that they were alone more than sufficient for his full compensation."[12]

Continuing his catalogue of laboring-class grievances against the Federalist Corporation, "A Mechanic" moved beyond the ordinances and their draconian enforcement to remind his fellow craftsmen of the Corporation's concerted attack on traditional artisan prerogatives by their continuous attempts to fix the price of labor, a policy that had, he noted, "created so much uneasiness that it was finally repealed by the legislature."[13] Summarizing the past six years of city government, he concluded that "the conduct of the Corporation throughout this business has been of a piece, and marks their character for aristocracy in lines too strong to be possibly mistaken." The remedy as he saw it was to vote against the Federalists, both in the Corporation and beyond, and to take public affairs into their own hands. "The remedy," he wrote, "is in your power."[14]

If the results of the 1796 election were any measure of success, the Democratic-Republican campaign among Philadelphia's laboring classes had borne plentiful fruit. The predominantly laboring-class districts of the Northern Liberties and Southwark voted Democratic-Republican by healthy margins and elected Blair McClenachan to Congress.[15] But of greater significance to contemporary party leaders was the election of John Swanwick to the city's congressional seat.[16] This Democratic-Republican inroad into traditionally "safe" Federalist territory was given the greatest importance by party leaders, who considered it their greatest victory that year.[17] Significantly, Swanwick's small plurality came from the five wards—New Market, Upper and Lower Delaware, North and South Mulberry—with the greatest proportion of workingmen. It was a pattern that would last well into the nineteenth century.

Yet while these gains were impressive and pointed toward the critical election of 1800, perhaps the most dramatic test of the new artisan Democratic-Republican political alliance came not in 1796, but in 1797 and 1798 as Israel Israel mounted two successive senatorial campaigns. Israel was a Philadelphia merchant who had made his fortune early, retired from commerce, and turned his energies to popular politics. He had won popular esteem for his tireless activities during the British occupation of Philadelphia, and especially for the privation he suffered as a British prisoner during the long winter of 1778–79. In the postwar era, his Cross-Keys Tavern became a frequent meeting-place for workingmen and opposition party leaders and held as good a claim as any to being the birthplace of laboring-class Democratic-Republicanism. In short, Israel was one of the most popular of the local Democratic-Republican leaders and one of the last, along with Thomas McKean, who could claim the mantle of revolutionary service. His relief work during the yellow-fever epidemic of 1793 only confirmed his popularity among the city's laboring classes.[18]

Israel Israel's political career began in 1793 when he entered the race for a vacant state senate seat. He lost that race to the Federalists, although he carried the laboring-class suburbs by a small majority. His next attempt at political office came in 1797 when, encouraged by Democratic-Republican victories the previous year, he again sought a state senate seat. Party leaders had chosen Israel as their candidate by a very "small majority" in September, and it was clear that his victory would hinge on the laboring-class vote. Israel's campaign began by addressing a persistent problem faced by all popular politicians in the post-Revolutionary era. A generation of political experience had taught workingmen to be wary of middling merchants and politicians who claimed to represent their interests, but who cared little about them after election day. This antipathy toward middling-class politics was reflected, in part, in the low rates of laboring-class political participation that the Democratic-Republicans were only beginning to overcome. It was with this lingering inertia in mind that "Freeman" exhorted the workingman who had abstained from politics not to "stand aloof when the good of [your] country urges [you] to step forward," for only "union and perseverance will give us the day."[19] The following day radical publicist Benjamin Franklin Bache made an even more direct appeal to the city's laboring classes by poking fun at the lingering deferential assumptions of the Federalists. Addressing "the great body of our citizens, the useful classes among us," Bache went to the heart of laboring-class sentiment, declaring that "the artisans and mechanics have too much respect for themselves to object to Israel Israel because he is not a merchant or lawyer."[20] Moreover, as Bache again reminded the city's laboring men, Israel had also taken a great "risk to his health in relieving the sufferings of his indigent fellow citizens [in 1793]."[21]

This last point had a special resonance because Israel's 1797 campaign was conducted in the midst of another outbreak of yellow fever. Less severe than the outbreak of 1793, the epidemic nevertheless drove the wealthy and middling classes from the city and left laboring-class families once again to shift for themselves. But unlike 1793, city workingmen now had their own organizations and the Democratic-Republicans to turn to for aid. At the outbreak of the fever, party leaders helped to create a tent city along the Schuylkill that allowed laboring-class families to flee from the crowded and fetid conditions they had faced four years before. A week later, Dr.

Michael Leib, the Democratic-Republican assemblyman from the Northern Liberties, introduced a bill in the state legislature to appropriate $20,000 "for relief of the distressed during the present calamity."[22] When the city's Federalist representatives joined their party's majority in the legislature to defeat Leib's bill, the Democratic-Republicans canvassed the wealthy and middling classes outside the city for contributions toward the running of the tent city.[23] Once again, Israel was prominent in the relief efforts, serving as one of the commissioners charged with distributing aid among the laboring classes.[24]

When election day arrived, Philadelphia's elite rode into the city from their country refuges, cast their ballots for Benjamin Morgan, Israel's Federalist opponent, and promptly returned to the safety of their estates. It was with shock and dismay that they received the election results: Israel Israel had won the election, carrying Philadelphia County and two wards within the city itself.[25] Reflecting on the election, Bache hailed Israel's election and the surprisingly close race the Democratic-Republicans had run in the traditionally Federalist city wards. The decisive factor had been Philadelphia workingmen, for despite the fact that "want of employment had driven away great numbers of the industrious classes [who] could not return to exercise their right of suffrage," the remaining workingmen had stood behind the Democratic-Republican ticket; the result, according to Bache, was that the Democrats had "trod closer on the heels of that opposed to it than for many years back."[26] As for Israel Israel, Bache celebrated his victory as a clear sign that the Democratic-Republicans were gaining strength in Philadelphia. Israel's election was especially important, Bache thought, because he was "one of those democrats most obnoxious to the faction; vice-president of the Democratic Society, an enemy to the prominent measures of the federal government . . . he is a plain man, of natural good sense, possessed of none of those brilliant attributes which a polished education gives."[27] He was, these attributes testified, a man of the people.

This was not the end to Israel Israel's long story, however. Even as he took his senate seat, the Federalist-controlled legislature moved against him. On December 13, 1797, the senate appointed a committee to investigate alleged irregularities in the October elections, and by February 1798, Israel learned that his election had been declared "illegal and invalid" by the Federalist majority in the senate.[28] In the special election held on February 22, 1798, Israel lost his seat to his October rival by the slim margin of 357 votes.[29]

Israel's defeat was not due to any lack of laboring-class support. The special election had drawn half of the city's adult males to the polls, making it the largest turnout in Philadelphia to that time.[30] Bache himself noted that "Israel stands higher now in proportion than he did in October" and took it as "incontestable proof that the Republican spirit of the city is rising."[31] Even the Democratic-Republicans' arch-enemy, William Cobbett, admitted to Israel's popularity, attributing his defeat to the large block of Federalist votes cast by traditionally non-voting Quakers.[32]

In the last analysis, Israel's defeat in the special election was less significant than the reaction of city Federalists, who feared the very prospect of a successful Democratic-Republican–laboring-class alliance. As the people of Philadelphia contemplated the meaning of the special election, city craftsmen began to reveal the pressure they had faced from their Federalist employers in the weeks leading up to the election. On February 26, an anonymous writer claimed that the Federalists had "threatened to deprive of bread those who are in their employ if they did not vote with them."[33] The next day "Brutus" reported several more examples of Federalist intimidation of laboring-class voters. In one case, "a *British* merchant went into Kensington the day preceding the election and there declared that no man should be permitted to work for him who would not vote for Mr. Morgan."[34] In another incident the carrot rather than the stick was offered: "one of the shipcarpenters in Kensington received a letter from a merchant in the city requesting him to exert himself on the day of election and make himself conspicuous and that he should have as much work as he could do."[35] And, "Brutus" reported, "in Southwark one *federalist* said he had a thousand cord of wood to haul and that no man should be employed to haul his wood who would not vote for Mr. Morgan."[36]

Nor did the Federalists stop at intimidation. Where the threat of unemployment was insufficient, the Federalists were also reported to have paid people's taxes, driven the poor to the polls in their coaches, threatened the custom of shopkeepers, and held the threat of eviction over the heads of tenants who voted for Israel.[37]

It is a testament to the courage of laboring-class Philadelphians that this Federalist pressure had little of its desired effect. As the results from the laboring-class districts demonstrated, workingmen were willing to risk their livelihoods and even their homes to maintain their independence. John Robinson, a city victualer, spoke for his fellow workingmen when he published his own rejoinder to Federalist intimidation: "This is to certify that James Read, inspector of flour of the Port of Philadelphia told me unless a man voted as he did, he should not be in his employment. I therefore hope that said Read will not offer me any more of his employment, or any other man of his disposition, as it will give me an opportunity to be employed by men of more liberal principles."[38]

The elections of 1796 and 1797–98 marked a turning point for the Democratic-Republican party and the laboring-class movement as well. After 1798 Philadelphia's Democratic-Republican leaders looked forward to the overthrow of Federalism with increasing confidence. The populist coalition of merchants, manufacturers, shopkeepers, and workingmen that they had envisioned since the beginning of the decade had come to pass and was more effective than they had imagined. "Will Mr. Adams say hereafter," one Democratic-Republican asked, "that we are not a divided people? [The Federalists] well know that tho' wealth may be on their side, nerve and fortitude, those revolutionary virtues, are on the side of the friends of liberty."[39] For their part, workingmen who had abandoned politics during the Federalist ascendancy and turned their attention to benevolent and trade organizations now returned, more powerful for the experience, to the political arena. A decade of internal organization had made them more united, conscious, independent, and articulate than at any time in the past, and these qualities would make them the focus of democratic politics for decades to come. One astute observer captured the essence of this new laboring-class

Democratic-Republicanism following the 1797 election: "Many of those who held elevated situations in the American revolutionary war have been more than once reproached with at least a partial abandonment of the principles for which they contended. The truth is, these persons have generally become rich [and an] overgrown rich man [cannot be] a good republican. It is remarkable that men of great wealth have generally in no age possessed for any length of time principles in favor of human rights; the people, and what is called the middle and lower classes have always been the guardians of this deposit, and with them it appears it will eventually remain. The great are continually fluctuating while what is placed low possesses stability."[40]

II

From the late 1790s, any account of Philadelphia's working class must recognize the complete and often subtle imbrication of the burgeoning laboring-class movement, anchored in the city's journeymen's trade societies and laboring-class neighborhoods, with the Democratic-Republican movement itself. The two movements developed in tandem and intersected each other in ways that, at times, made them nearly indistinguishable. Even at its moments of greatest independence, as in the 1805 cordwainers' strike, the laboring-class movement always functioned in a milieu of Democratic-Republican politics.

It proved to be a creative association for both movements. Over the course of the next two decades, the Democratic-Republican party was forced by the powerful presence of the craftsmen in its ranks to wrestle with a range of issues—the most important of which was the meaning of democracy in an increasingly class-divided society—that would place it at the forefront of an increasingly modern style of politics. The laboring-class movement was, however, the greatest beneficiary of the interchange, for the Democratic-Republican alliance forced Philadelphia artisans to re-examine their small-producer tradition in the light of a new set of ideas espoused by the radical leaders of the party. It was from this compound of artisan moral traditions and radical strands of thought that the fundamental ideas of the American working class would take their earliest form.

The closing years of the eighteenth century were crucial ones for laboring-class Philadelphians. In politics the Democratic coalition survived the Federalist resurgence of 1798–99, the threat of war with France, and passage of the Alien and Sedition acts, which sought to silence the opposition press and disenfranchise the radically anti-Federalist Irish who flocked to Philadelphia after 1785. By 1799, the opposition movement was powerful enough to elect Thomas McKean the first Democratic governor of Pennsylvania, and in the 1800 election Jefferson and the Democratic-Republicans carried Philadelphia by overwhelming majorities.[41] The year 1800 also marked the final triumph of Israel Israel, who rode the Jeffersonian upsurge to victory as sheriff of Philadelphia.

A good measure of the credit for the Democratic-Republican triumph went to a young newspaper editor and the laboring-class wing of the party he had helped to organize. William Duane was born on the rural frontier of upstate New York in 1760. When his father died in 1765, Anastasia Duane left the family farm near Lake Champlain and moved first to Philadelphia and then to her native Ireland, where Duane learned the printing trade in the small village of Clonmel, Tipperary.[42] As a young man, Duane served as parliamentary reporter for the London *General Advertiser*, working there until 1786, when he sailed to India to undertake an appointment as clerk to the East India Company. It was there, in 1789, that Duane began editing the *Bengal Journal*. It was a fateful year for both France and William Duane, and the ardent young republican soon became a thorn in the side of the British colonial administration. By 1794, Duane found himself again bound for England, having been deported for his pro-French public views. Once in London, Duane quickly undertook the editorship of *The Telegraph*, a newspaper renowned less for its circulation figures than for its support of the radical London Corresponding Society.[43] The offices of the *Telegraph* often served as a meeting-place for the Corresponding Society, and Duane rapidly became involved in its activities.[44] When the corresponding societies were eventually driven underground by a frightened and vindictive Pitt, Duane fled to his native country to avoid imprisonment for his Jacobin ideas.

His radical credentials secure, in October 1796 Duane arrived in Philadelphia, where he served as editor to two small city newspapers before finding more permanent employment on Bache's *Aurora* in 1798. Again, fate turned Duane's way, and shortly after Bache succumbed to yellow fever in September, his widow named William Duane the new editor of the family newspaper. Margaret Bache could have made no more fortuitous choice. To his editorship of the *Aurora* Duane brought a first-hand experience of British oppression and a close understanding of the radical ideas that were sweeping through contemporary England, Ireland, and France. These qualities joined in the pages of the *Aurora* and made it the nation's principal arena for the encounter of European and American radicalism.

In assuming editorship of the *Aurora*, Duane found the radical path well laid. Since its inception in 1790, the *Aurora* had been a vehicle for radical opinion as well as the semi-official organ of the Democratic-Republican party.[45] Bache advertised himself as the Philadelphia agent for Daniel Issac Eaton, the London bookseller who spent the greater part of his adult life in English prisons for printing and selling radical tracts, Paine's *Rights of Man* foremost among them.[46] Like most early American newspaper ventures, the offices of the *Aurora* doubled as a bookstore and in a typical week Bache offered for sale the works of Algernon Sidney, Paine's *Rights of Man*, Volney's *Works*, *The Works and Speeches* of Arthur O'Connor, More's *Utopia*, Pigott's *Political Dictionary*, and *Monarchy, No Creature of God's Making* by Judge Cooke "in the time of Oliver Cromwell."[47] Bache was an internationalist as well, and the *Aurora* carried news of the rise and repression of English radical reform movements, the progress of the French Republic, and the mounting discontent in Ireland. Arguably the foremost native-born spokesman for Irish rights, Bache portrayed the United Irish revolt as the child of America's own revolution, writing with more enthusiasm than grace, "O America remember that thy revolution has actually kindled the fire of liberty in the old world."[48]

Although, as a Democratic-Republican journal, the *Aurora* always printed a wide variety of opinion, Bache's radicalism spoke to a growing segment of the Democratic-Republican coalition. This segment, composed of radical professionals, shopkeepers, and working-men, would form the foundation of laboring-class politics in Philadelphia. Significantly, this segment also contained a large Irish and Anglo-Irish component, reflecting the resumption of Irish immigration following the Revolutionary war. [49]

William Duane was well placed to take up Bache's mantle as the leading propagandist for a radical brand of Democratic-Republicanism. Despite his American birth, his Irish ancestry and his widely publicized experiences in the British empire caused many, Federalists and Democratic-Republicans alike, to think of him as Irish-born. Indeed Duane was so widely thought of as an Irish immigrant that in 1799 he took out naturalization papers in order to protect himself against deportation under the Federalist Alien Act. [50] But whatever his ancestry, Duane's sympathy for the United Irish movement and his defense of Philadelphia's Irish community placed him, and the *Aurora*, at the center of laboring-class politics in Philadelphia. [51]

Duane's popularity among the city's radical laboring and middle classes was the result of his democratic radicalism, his Irish sympathies, and, after 1799, his reputation as a popular hero. In 1799, a volunteer troop of young Philadelphia Federalists marched into rural Northampton County to put down Fries Rebellion. As word reached Philadelphia of the troops' excessive brutality and flagrant violation of civil rights, Duane printed the details of their depredations in the *Aurora*. For this, the officers of the troop pulled Duane from his office, dragged him into Market Street, and nearly beat him to death. [52] His stand for the principle of a free press, the rights of man, and against the authoritarianism of the Federalists transformed his actions into the elements of a popular legend.

Duane's greatest contribution to Philadelphia's laboring-class movement, however, came less from his popularity than from his propaganda for new ideas and from his talents as an acute and critical analyst of the changing world of the city's capitalist economy. Two themes animated Duane's thought throughout his long career. On the one hand he combined the doctrines of radical political economy with the major tenets of the small producer tradition to forge a critique of the declining position of the producer in American society. On the other hand, he labored tirelessly to inform Americans about the drift of English industrial society and the degrading conditions that it brought to farmers and mechanics there. It was only by understanding the nature of English development, Duane believed, that the same mistakes could be avoided in America and that an industrial society could be built which balanced the needs of small producers with the imperatives of national development.

Duane first announced these themes in two articles published in the spring of 1800. In the first, he drew upon a central tenet of the small-producer tradition to criticize a Federalist plan to create a national debt. "We consider prosperity to imply competency without excess," he declared, "sufficient for all public purposes and for private support."[53] A national debt, he argued, would only threaten the competency of artisans and farmers in order to make wealthy men even more affluent. A month later, he began his comparison of Anglo-American events by likening Federalist prosecutions for libel and sedition during the 1790s to the prosecution of John Wilkes under similar statutes in Britain thirty years before.[54]

These twin themes would animate all of Duane's editorials, whether he was discussing the issue of banks, civil liberties, or the encouragement of domestic manufactures and internal improvements. Taken together, the combination of small-producer values and English social criticism that they represented would prove vital to the modernization of artisan political culture in the early years of the nineteenth century. Nowhere was Duane's contribution to this new political culture more evident than in a series of essays he wrote between January and March 1807 under the collective title "Politics for Farmers and Mechanics." Like all of his work, these essays were dedicated to small producers in general, whether farmers or mechanics, because, as Duane put it, "the importance and interests of actual industry are common to the whole body of industrious men who are not above the *dull pursuits of civil life.*"[55]

The opening essay began with a warning to the productive classes that some members of the community desired to "bow down and rise upon the necks of their fellow citizens over whom they fancy they possess either greater talents or greater riches." If allowed to gain power, he predicted, these men would act like "the profligate *Arnold*, [turn] their backs upon virtue, lay claim to honor while playing the knave and end with becoming a sore on society and a disgrace to human nature."[56] Here, in a revealing analogy, Duane compared non-producers to Benedict Arnold, the Revolutionary Army general who took command of Philadelphia following the British evacuation in 1778. Haughty and ambitious, Arnold carried the aristocratic culture of the British occupation into the post-occupation era and in 1779 married Peggy Shippen, daughter of one of Philadelphia's wealthiest Tory families. Arnold earned the antipathy of the city's laboring classes by hosting lavish parties for Philadelphia's Tory elite at a time when ordinary patriot families hovered near the edge of subsistence. When he betrayed the American cause and defected to the British in 1780, Arnold became a symbol for Philadelphia workingmen, representing the arrogance of wealth and power as well as the self-serving ethos of the city's aristocratic elite. Duane's use of Arnold as a metaphor for Philadelphia's parasitic non-producers provides striking evidence of the continuity of working-class culture from the Revolution to the second decade of the nineteenth century.[57]

Against such unbridled acquisitiveness and lack of community concern, Duane reasserted the communitarianism of small producers. The "ends for which civil society is instituted," he recalled, is "the promotion of the happiness of the whole or of the greatest number," and as "farmers and those who acquire support from labor" are 17/20ths of the population "that policy which secures the *happiness* of that majority must prevail."[58] All other interests must be subordinated to it. Here, for those readers who could remember 1779, was the argument of the Committee of Thirteen wrapped up in utilitarian garb.

Writing from the vantage point of Philadelphia's artisan community, Duane went on to anatomize the ideal society embodied in the small-producer tradition. The key to that society was the utility of labor and Duane called upon the authority of Benjamin Franklin to underline his point. "The earth and the waters are the sources from which all true riches are produced," Franklin had written, "but the earth and the waters would be unproductive without *labor*, therefore the *labor of tillage* is the first, and the *labor of manufacturers* the second means of acquiring national and individual wealth."[59] But while labor was central to any society, Duane also recognized that a community of producers needed to support commerce, but only if that commerce was limited to the exchange of American produce for those foreign goods which the new nation lacked or was unable to produce. "*Agriculture* and *commerce* we hold to be *inseparable*," Duane told his readers. "But let us separate good from evil, let us set the fair trade on its right foundation, let us not involve it with the rash speculations of every adventurer who may be cast on our shores."[60] The "proper legitimate bounds" of commerce, Duane argued, was the "*fair* commerce of the productions of the *American soil*."[61]

To Duane, it was those merchants who engaged in the peacetime carrying trade and those who garnered enormous profits from provisioning contracts in time of war who were subverting the producers of the Republic. As one writer put it, in the republican idiom of the educated middle classes, "it has been the commercial interest of every nation . . . which has prepared the evils that have hastened their ruin. Aristocracy and oligarchy seem the natural result of commerce for . . . they destroy that simplicity and equality of manners essential to equal rights and generate a pride and ostentation which claims distinctions and classes."[62] Duane concurred, but moving beyond classical republicanism to the logic of the small producer tradition, he also warned workingmen that "every *idle* fellow . . . *above the dull pursuits of civil life* . . . expected to live on the sweat, and the labor, and the industry, and the talents of the virtuous part of the nation."[63] In Philadelphia, he claimed, the city's ruling class of merchant capitalists meant to make the city's workingmen "the mere tributaries of the more wealthy part of the city."[64]

Duane went on to remind his urban readers of the intrinsic value of their labor and the "self-respect which every *mechanic* should feel as forming part of that great basis upon which society is erected and without which society could not exist in a social and happy order."[65]

Then, moving beyond the small-producer tradition, Duane introduced the ideas of the radical economic critics of the English industrial revolution.[66] Duane introduced these new ideas in the course of an analysis of the misdirection that English development had taken since the seventeenth-century civil war.

The evils that had befallen the farmers and mechanics of Britain could be traced, Duane argued, to the dominance of commercial and industrial interests in the setting of state policy following the Napoleonic Wars. The fruits of that policy, "the *paper money*, the *mercantile*, and the *banking* system" had left "the poor and unoffending farmer . . . the slave and *vassal* of his noble or ignoble landlord." "Crushed by taxes," the English producer was "sent in his old age to the poor house, his children scattered over the world to fight the *mercantile battles* of their despots." "Such would be your fate," Duane warned his readers, "if British policies prevailed [for] like the *yeomanry* of Britain you would sink to destruction."[67]

Here, as in many of his "Letters," Duane's targets were Federalist merchant capitalists who sought to transplant British policies into domestic soil. As in England, the interests of commerce would inevitably lead Americans into commercial wars and violate the domestic peace under which small producers flourished. It was always "the FARMERS and industrious classes [who] PAY AND FIGHT ALL!" Duane argued, while merchants "accumulate *thousands* by their foreign traffic or by *war contracts*." Moreover, he noted, it was these same merchants who were "comparatively exempted from the calls of the *tax gatherer* and are never to be seen in the ranks of militia."[68] Here Duane spoke directly to the memories of Philadelphia craftsmen, for during the Revolution they had watched engrossing and profiteering merchants make enormous fortunes at their expense and then use their new-found wealth to pay for militia substitutes.[69]

If an overblown commerce had led England toward involvement in international wars, with grotesque and murderous consequences for British working people, Duane laid a good part of the blame on the nation's farmers and mechanics themselves. "The cause [of the English producer's plight], he argued, was that the people who *fight*, *suffer*, and *pay* for all wars suffered the power to be wrested from their hands. The interests of their peace and happiness they trusted for *too long a time* to deputies and the people became supine from the want of a prompt and adequate control."[70] The lesson to be drawn was clear. If American farmers and mechanics wished to avoid following their British counterparts down the road to subjugation, they must act together against those who were determined to undermine their liberties. The solution, Duane declared, was a revival of the Revolutionary spirit. "Had it not been for the Revolution," he wrote, "we should have been surrounded by *tithes*, by *privileged colleges*, by privileged *church rules* and church wardens, visitations, and *taxes*, by a privileged *clergy* and *Nobles of Nova Scotia*."[71] "Did the American Revolution ever intend," he went on to ask, "that because a man had not riches he was to be deprived of liberty also?" Did it decree "that the rich man was to be his keeper?"[72] No, answered Duane, because in America "the *vulgar*, the *people* will always be ready to defend their liberties against those *men of birth*" who would make the producers' "tears and the tears of your hungry children, the sweat of your brows, support their idleness."[73]

If war was, for Duane, the necessary outcome of an aggressive foreign commerce, taxation was the means by which the degradation of productive labor was carried out. Duane viewed domestic taxation as the instrument of a commercial ruling class, which used it to finance both its privileged place in society and the necessary paraphernalia of empire. "Taxes, taxes," he wrote, "these are among the glorious consequences of constitutional fortifications, navies [and] standing armies."[74] It was current Federalist proposals for just such a system—an enlarged navy, a standing army, and military fortifications—to be paid from internal taxation that caused such immediate alarm. In Duane's mind, the United States was fast duplicating the errors of England, and he again warned America's small producers that "if you once submit and give way to internal taxation, unless for the promotion and improvement in your interior, those who will live on your industry in collecting taxes will soon ruin you with expenses."[75]

Duane was neither an original nor a systematic thinker, yet his essays were critical in the development of a specifically working-class outlook in Philadelphia. At a time when laboring-class attention was most often focused on the immediate effects of economic change and the growing division between employing masters and journeymen, Duane pointed toward a theory of exploitation and reasserted the necessity of political action in gaining small producer objectives. Duane's message to Philadelphia workingmen recalled their small producer values and urged them to reclaim their pivotal position in the production process. If they ignored this duty, he warned, the way would be open for merchant capitalists, currency manipulators, and land speculators to control their livelihoods and to dictate local and national policies. In the end, that would lead them, like their English counterparts, not toward a life of independence and competency, but into a life of dependence, degradation, and slavery.[76]

The importance of Duane's propaganda and agitation efforts on behalf of city craftsmen cannot be over-emphasized. In the decade and a half after he undertook responsibility for the *Aurora*, Duane proposed to Philadelphia artisans that they could best represent their own interests, and the common interest of the community, by a combination of political and trade action. His prescription for the illness of the time was nothing less than a movement of artisans and workingmen to reclaim the rights of a democratic community regulated by "rational labor, temperance, and domestic love."[77]

Duane's program followed a long-established Philadelphia tradition that had called workingmen to political action since the early eighteenth century. But there were also crucial differences. Eighteenth-century political leaders had roused the city's craftsmen to action as a counterweight against the power of a mercantile and proprietary elite with marked aristocratic pretensions. Duane, on the other hand, viewed craftsmen as the sole producers and natural legislators of the community itself. His argument was that of the revolutionary Committee of Thirteen, sophisticated by deep contact with English and French radical thought. But while essential, this theoretical distinction led to a second, and even more critical, difference in Duane's program: his support of the city's growing journeymen's movement. Of the large and influential community of expatriate radicals in Philadelphia, his was the only voice of support for the journeymen shoemakers in their 1805 strike and subsequent conspiracy trial.[78]

The strike and trial of the journeymen shoemakers in the winter of 1805–6 is well known, having attracted the attention of labor and legal historians since John R. Commons first wrote about them in 1909.[79] Briefly, the journeymen shoemakers organized a strike against their employers when the latter sought to reduce their wages during the slack winter months. In response, the leaders of the journeymen's society were jailed, charged with common-law conspiracy to restrain trade, and stood trial in March 1806. The trial was at once a contest between journeymen and employers, critics and advocates of English common law, and the merchant-manufacturer and laboring-class factions of the Democratic-Republican party. In the end the journeymen were found guilty and made to pay a small, symbolic fine while restraint of trade was established as a legal precedent and made available to employers as a weapon for use against other journeymen's societies.

Duane's defense of the journeymen began immediately after the arrest of their leaders in November 1805. In "The Price of Labor," he deplored the arrest, both for its violation of principle and for its probable consequences. "Is there any power," he asked, "that can lawfully and constitutionally determine the price of a man's labor to be less than what he chooses to accept voluntarily and of his free choice?" If such a power existed, he declared, then "the constitution is a farce and the bill of rights is only a satire upon human credulity."[80]

More ominous, however, were the results which Duane foresaw from the suppression of the journeymen's society. With the recently enacted English Combination Acts of 1799–1800 fresh in his mind, Duane speculated on the true nature of the arrests. "This no doubt is another of the glorious *shoots* of English *common law*—and no doubt as the *doors of suffrage* are already too wide and as the 'peasant comes too close to the heel of the *courtier* it galls his kibe,' the doors of industry are to be closed so that a breed of *white slaves* may be nursed up in poverty to take the place of the *blacks* upon their emancipation."[81] Consistent with his comparative perspective on Anglo-

American social development, Duane foresaw the common-law suppression of the journeymen as leading to a society in which "poverty, stupidity, disease, and vice have superseded industry, intelligence, health, and innocence."[82] The arrest of the journeymen cordwainers was but another step in the growing Anglicization of American society.

In his initial coverage of the trial itself, Duane promised his readers that he would follow the journeymen's progress "again—and again."[83] He did so diligently. The day following the appearance of "The Price of Labor," Duane printed the lengthy "Address of the Working Shoemakers," which explained the journeymen's predicament to the public and gave their reasons for joining together to redress their grievances. "The name of freedom is but a shadow," they wrote, "if for doing what the laws of God and the laws of our country authorize, we are to have taskmasters to measure out our pittance of subsistence."[84] They promised their fellow citizens that they would face "the oppression to which we are exposed with constancy and the temper that befits men actuated only by justice and the spirit of freemen."[85]

When the journeymen were found guilty on March 29, 1806, Duane recorded the defeat in the clearest of class terms. "We must religiously abhor the doctrine which makes it *conspiracy* for *poor men* to regulate the *value of their own labor*, and that protects rich men who conspire and say for what sum a poor man shall sell his labor or services."[86] "As it has been declared to be a conspiracy for *journeymen* to meet and regulate the prices for which they will work," he went on to ask, "what are we to call the *combination of masters* to make the journeymen work for lower wages than they think themselves entitled to?"[87] Finally, Duane noted the class nature of the Mayor's Court in which the journeymen had been tried. Anyone innocent of knowledge about the legal system, he declared, "would unquestionably have concluded that Mr. Moses Levy [the presiding judge] had been paid by the master shoe-makers for his discourse in the court; never did we hear a charge to a jury delivered in a more prejudiced and partial manner." "From such courts, recorders, and juries," he implored, "good lord deliver us."[88]

The light fines given to the journeymen did little to diminish their cause in the public mind. Among the Fourth of July toasts offered that year was one honoring the journeymen's defense attorney, Caesar A. Rodney, for remaining "honest and true to the principles of '76."[89] This was followed by a toast to the shoemakers themselves: "To the *Sons of St. Crispin*, the second *mechanical* martyrs under the present reign of terror. May the mechanics of America duly appreciate by their fate the designs and wishes of the advocates of the common law of Britain and learn to consider its advocates as the common enemy of human happiness and independence."[90] Duane's writing had found its mark.

Duane carried the cause of the journeymen shoemakers into the following year. In the fourth installment of his "Politics for Mechanics," he cautioned American craftsmen against sharing the fate of their English counterparts which he called "perhaps as stupendous a monument of *national slavery* as has existed from the earliest epochs of time."[91] The city aristocrats were in high spirits because of the shoemakers' conviction, he wrote, and as a result were declaring "now we will manage these *porters* and *draymen* and the *mechanics*—these men earn too much—if they do not earn half as much it would be the better for us." "Remember the case of the shoemakers," Duane urged, for "the case of the shoemakers is only another proof of the tendency of *lazy luxury* to enslave the men of industry who acquire their bread by labor."[92] Underlining the small producer values embodied in the journeymen's cause, Duane counseled city mechanics to "watch every encroachment on the price of your labor, tell [your employers] you are entitled to independence as well as themselves."[93]

In Duane's mind the case of the journeymen shoemakers represented the beginnings of a new form of class conflict in America. Reviewing the sad story of the English handloom weavers who "dare[d] not to meet to regulate the price of their own labor, even when it was too little to buy bread," Duane drove his point home with a warning to the mechanics of America. "The case of the *shoemakers*," he predicted, "is a *precedent* set up in advance to prepare you for such a fate."[94] The drift of work, politics, and social life in Philadelphia was indeed taking on an increasingly class-inflected character in the decade and a half before the second Anglo-American war, and though he continued to hold to the idea of a balanced community of producers, Duane recognized the growing saliency of class divisions in Philadelphia life and continued to support "the check apron women and the leather apron men" of America throughout his long career.[95]

The fruits of his campaign to galvanize the city's working class would take a generation to ripen fully, but the immediate impact of Duane's arguments on the minds of Philadelphia workingmen can be measured by examining the social origins of the Democratic candidates for the city's common council. In ordinary times laboring-class Philadelphians were always more concerned with local than with state or national affairs, and the election of the common council, which legislated on those issues closest to the lives of workingmen, was a close indicator of laboring-class political sentiment.

Prosperous artisans and manufacturers dominated Democratic-Republican council tickets from the beginning, but in 1801 Philadelphia merchants claimed a full 40 percent of the council nominations, guaranteeing them a substantial voice in all council affairs.[96] Following the publication of Duane's "Farmers and Mechanics" essays, however, merchant representation fell precipitously; in 1807 they claimed less than 17 percent of the nominations, and by 1812 local merchants composed a mere 12 percent of the council candidates. While it is unlikely that Duane's efforts were the sole cause of artisan dominance of the Democratic ticket, the timing of the merchant decline suggests that he played a pivotal role.

While the make-up of the common council candidates moved ever closer to Duane's ideal of small producer representation, a more direct measure of his influence came from the journeymen cordwainers themselves. During the years following their 1806 trial, the journeymen paid tribute to Duane and his principles by organizing themselves as the Democratic Cordwainers of the City and County of Philadelphia.[97] Shortly thereafter, on the eve of the 1812 election, the cordwainers moved beyond their trade society concerns to call upon "all the free and patriotic journeymen shoemakers . . . of Philadelphia, all who are adverse to the reign of terror established by our Tory councils, all who recollect the fines and imprisonment of the *craft* by Tory judges and Tory juries for asserting the rights of freemen, and all who are desirous of preventing a repetition of such conduct and such scenes . . . to attend and *show* themselves to be the friends of liberty and independence" by voting the Democratic ticket.[98]

To drive home their point, the shoemakers organized an election-day march that paraded through the laboring-class districts of the city. Led by George Alcorn, a journeyman shoemaker, the Democratic Cordwainers carried aloft a banner that answered the appeal for combined trade and political action that Duane had made in his "Politics for Mechanics." Beneath the emblems of their trade the cordwainers had emblazoned: "DEMOCRATS, UNITE AND CONQUER."[99] In the 1812 election the city's working classes did just that, sweeping Democrats into virtually every city and state office. For Duane, it was a fitting reward for his labors.

III

Duane's advocacy of the nation's producing classes and his admonitions about the increasingly British drift of American society came at a time when economic life in Philadelphia was beginning to follow the path of capitalist industrialization marked out in England a generation earlier.

During the eighteenth century, Philadelphia had been a classic entrepôt, a commercial city that received and processed grain, hides, and wood from its hinterland and imported English manufactures for home consumption in return. But by 1830, Philadelphia was already an industrial city. A mere half-century after the Revolution, the compact walking city of 39,000 had grown into a bustling metropolis of 185,000 inhabitants.[100] Where merchants, artisans, and shopkeepers had once lived side by side in a city that stretched less than a dozen blocks from the Delaware waterfront, after 1820 they lived in separate ethnic and class-segregated neighborhoods that extended west across the Schuylkill River and several miles north and south of the city center.

The signs of economic change were apparent everywhere. Commerce had been in decline since the turn of the nineteenth century, when Philadelphia's West Indian and southern European markets waned and New York became the favored port of entry for foreign commerce.[101] This deterioration in Philadelphia's commercial position can be seen most clearly in the statistics of foreign trade. In the final decade of the eighteenth century Philadelphia accounted for nearly a fifth of all U.S. trade, but only two decades later the city's maritime traffic had fallen to little more than a tenth of the national total. By the 1830s Philadelphia exports amounted to a mere 4 percent of America's foreign commerce.[102] Although commerce would remain a source of substantial employment in the early industrial period—fully 20 percent of the city's adult males continued to follow commercial occupations as late as 1820—manufacturing employment gained increasingly after the embargo of 1807–9.[103]

This transformation of Philadelphia into a manufacturing center was already apparent in 1788 when the French traveler Brissot de Warville noted that "manufactories are rising in the town and country, and industry and emulation increase with great rapidity."[104] Tench Coxe, Assistant Secretary of the Treasury and one of America's foremost advocates of domestic manufacturing, confirmed Warville's observation in a brief census of Philadelphia manufactures that he published in 1804. In the closing years of the eighteenth century, Coxe found that the old commercial city of former years was fast being transformed by manufacturing establishments making hats, buttons, thread, lace, pottery, and gold and silver wire. Moreover in Germantown, a suburb northwest of the city, Coxe discovered a virtual army of German immigrant stockingmakers industriously laboring over their imported stocking frames. And in the Northern Liberties, only a few miles from the center of Philadelphia, he marveled at the Globe textile mill, which housed not only the latest carding machines but Arkwright spinning frames and 120-spindle mechanical mules.[105]

As the statements of Warville and Coxe reveal, in the quarter-century following the Revolution, Philadelphians witnessed the beginnings of an industrial revolution. It was an experimental period for everyone involved, as much for the merchants and master craftsmen who became industrial capitalists and created the city's outwork and manufactory systems as for the journeymen and half-trained apprentices who labored under their control. The uncertainty of the era was underscored by those who styled themselves the "manufacturing interest" of the city when, in 1787, they formed the Pennsylvania Society for the Encouragement of Manufactures and the Useful Arts in the hope that an emulation of English industrial machinery would provide a safe and easy way to transform Philadelphia into a cornucopia of industrial wealth.[106] These early attempts at industrial emulation failed at least as often as they succeeded, as the story of John Nicholson's ill-fated Roxborough manufacturing village and the public sale of failed enterprises by the Philadelphia sheriff amply testify.[107] But despite numerous setbacks and failures, the idea of domestic manufacturing persisted and became, in the early years of the nineteenth century, a mythic prescription for national well-being.[108]

If it had not been for the Napoleonic Wars and Britain's refusal to accept American neutrality, however, Philadelphia might still have become a declining port city struggling to turn urban manufacturing into a compensatory source of income. The possibility of such a fate was all too real in the post-Revolutionary era, for by the mid-1780s serious competition came not only from New York City but from the rapid growth of Baltimore, whose geographical position gave it privileged access to Pennsylvania's rich Susquehanna River valley.[109]

It was, then, the anti-British embargo of 1807–9 and the subsequent second Anglo-American War that propelled Philadelphia's transformation into a manufacturing center. The lack of competition from British imported goods between 1807 and 1809, and again from 1812 to 1815, provided a natural protective tariff for domestic production at the same time that idle merchant capital sought new forms of investment. By the end of 1808, the *Aurora* celebrated the rise of domestic industry by publishing a list of city manufactories that produced goods formerly imported from England. Under the protection of Jefferson's embargo, the paper reported, Philadelphia supported six textile, four glass, three chemical, two shot, and one each soap, lead, and earthenware manufactories.[110] On November

17 of that same year the elite Merchants and Manufacturers Society met at the first of a series of annual dinners held to congratulate themselves for the leading role they were playing in Philadelphia's ongoing industrial transformation.[111]

While these official and semi-official accounts point to a city in the midst of industrial change, it was Tench Coxe's 1810 industrial census that provided the first sustained portrait of early industrialization in Philadelphia. Compiled as part of the nation's first industrial census, Coxe's report revealed a growing number of substantial manufactories scattered among the city's smaller shops and confirmed Philadelphia as the national center of industrial production.[112] This mixture of manufactories, small shops, and outworker's homes was most apparent in the production of cotton cloth, where six manufactories produced half as much as all of the city's other cloth workers combined.[113] At the same time the city supported twenty naileries whose yearly production amounted to four million pounds and 201 blacksmiths who crafted everything from door bolts to hinges for domestic consumption and the export market. Among other metalworking establishments, Coxe's survey counted ten gun manufactories, 28 cutlers, and 42 copper, brass, and tin manufactories.[114]

Like metalworking, the production of leather goods quickly moved away from the independent artisan shop, and by 1810 Philadelphia's 59 tanneries produced hides worth $750,000. Taking these hides to their rented garrets and row houses, the boot and shoe makers—already the city's largest group of outworkers—supplied local as well as newly opened southern and western wholesale markets with shoddy goods worth more than $1.7 million.[115]

Consumer goods, too, showed a general shift toward larger-scale production. To clean its citizens and light their homes and workplaces the city's 22 soap and candle manufactories produced over 3.3 million pounds of soap and nearly 1.5 million candles. Cabinetmaking, long both an art and a craft in Philadelphia, was divided among the city's 67 shops, where masters and journeymen annually produced furniture valued at $456,000.[116]

Pleasure was also an integral part of the city economy and Philadelphia's 18 distilleries and numerous breweries produced 1.3 million gallons of spirits and 56,000 barrels of beer each year. Not to be outdone, the city's cigar makers, working at home or in small garret shops, rolled more than 31 million cigars for American smokers.[117]

Although modern statisticians could easily find fault with Coxe's survey, the portrait it paints of Philadelphia's early industrialization is an important one. Compared with the city he surveyed a decade before, Philadelphia had been transformed. Less tied to commerce than at any time in its history, by 1810 Philadelphia displayed all the characteristics of a developing industrial economy.

This picture of early industrial Philadelphia was confirmed a decade later by the industrial returns of the 1820 census. Then, as in 1810, textiles, leather goods, and metalworking dominated Philadelphia returns. In the fall of 1819, when the returns were gathered, the census marshals counted 30 textile, 25 leather, and 15 metalworking factories in the city and county. Next in rank were the city's ten flour mills followed by numerous small factories producing construction and consumer goods.[118] Likewise, a walk through the city late in 1819 would have revealed five furniture manufactories, an equal number of carriage and wagon makers, and six each hat, brick, soap and candle, and paper-making firms.[119] Considering that the 1820 census was compiled at the depth of the worst depression in Philadelphia's history, the number and variety of manufacturing firms operating in the city is eloquent testimony to the rapid progress of the city's industrial transformation.[120]

Visitors to Philadelphia in the early industrial era were often impressed by the large mills and semi-automated factories that were found there. But as dramatic as these factories were, it is important to remember that the scale of production in Philadelphia remained small before the Civil War. Of all workers recorded by the 1820 census, nearly two-thirds worked in shops employing fewer than six people and only one in ten worked with nine or more coworkers.[121] In fact, only 0.4 percent worked in factories employing 20 or more people. Thus in Philadelphia, as in England and France, early industrialization and the factory system were not synonymous. The transformation of Philadelphia from a commercial to an industrial city depended not upon the concentration of machinery and large numbers of workers under one roof, but upon a more subtle transformation which subordinated the pre-industrial craft system to the will of rising industrial capitalists.

In order to take command of the city's rapidly developing economy, early manufacturers realized that they had first to enter and control the production process itself. As they quickly discovered, however, this would be no simple or easy task. In those crafts where production continued to rely upon traditional skills, as in cabinet-making and carpentry, the craftsmen held a virtual monopoly on the skills required to transform raw materials into finished goods available for market. What troubled the city's prospective employers was that this monopoly of skill not only allowed craftsmen to demand high wages but permitted them to retain control over the labor process as well. So long as skill remained in the hands of producers, manufacturers came to realize, they were all but helpless in their attempts to set production quotas, product standards, or even the length of the working day. Until competition and mechanization undercut this monopoly of skill, the authority of the nascent manufacturer was effectively limited, and he was forced to devise new ways to control production and gain control of his workmen.

With its problems of employee discipline and production control, this indirect path toward industrial development attracted mostly small merchants and entrepreneurial masters, men with modest capital but an understanding of the production and marketing process. Larger manufacturers, who sought complete control of both workplace and worker, eschewed the outwork and central shop system of their smaller comrades, and concentrated their energies instead on new machine-driven production processes. By avoiding traditional craft-centered techniques and organizing production on an entirely novel basis, the large manufacturer was able to employ a less skilled work force and thus pay lower wages while maintaining control of the production process from the start.

In Philadelphia, machinofacture, as this process came to be known, first took root in the textile industry where employers used the cheap and more tractable labor of women and children to card wool and cotton and spin the yarn that would be woven into cloth by the city's handloom weavers.[122] This process is borne out by the 1820 census of manufacturers, which gives us our first view of the nature of women's and children's employment in the city. At the beginning of the third decade of the nineteenth century, women and children made up a majority of the manufacturing labor force, together comprising 52 percent of the city's industrial working class. Befitting the uneven nature of early industrialization in Philadelphia, most women and children worked in the smaller shops of the city. Thus 66 percent of the women and 70 percent of the children counted by the census worked in shops with five or fewer employees. This contrasted with 56 percent of the men who worked in establishments of that size. The largest firms, those with 20 or more employees, hired equal numbers of men and children, but less than half as many women.[123]

Not surprisingly, women were concentrated in textile production, which alone accounted for 51 percent of all women employed in manufacturing occupations. After textiles, women were employed in a variety of industries: 18 percent made leather goods, 10 percent worked in the hatmaking and paper-product industries of the city, while 11 percent worked to produce the assortment of light consumer goods that were transforming everyday life in the expanding nation. Children's employment followed a similar pattern, with 23 percent employed in textile production, 15 percent in leather goods, and 3 to 6 percent in each of the city's other industries. In fact, the only important difference between the employment of women and children in 1820 was that while children were found in every branch of manufacturing, women were excluded from the metalworking, flour milling, book binding, carriage making, and, interestingly, the soap and candlemaking industries. The absence of women in soap and candlemaking, long a traditional female domestic task, underlines the dramatic decline of home production in nineteenth-century America and confirms the existence of a wide-ranging consumer market throughout the nation as a whole.[124]

For their part, men were found in every firm surveyed. Like their female counterparts, the greatest concentration of adult men was in textiles, where they made up 21 percent of the male industrial labor force. Beyond the clothing trades, 18 percent of Philadelphia male industrial workers labored in leather goods, 11 percent in metal-working, and 7 percent in flour milling. The remaining 43 percent counted by the census were evenly divided among the city's other industries, each of which accounted for 4 percent of male employment.

Together these figures reveal a city enmeshed in the process of industrial transformation. The diversified and well-developed manufacturing base, the production of low-cost consumer goods, the beginnings of capital-goods production in the metal and machine-making shops of the city, the employment of women and children in the place of skilled male labor—all of these indicate the changing nature of the social relations of production in early nineteenth-century Philadelphia.[125] The small independent producer would continue to play an important role in the economy, but as the century wore on, he would find himself increasingly forced to depend on wages for subsistence and others for employment. This trend, which first appeared in the Revolutionary era, gained momentum after the turn of the nineteenth century and with each succeeding decade came increasingly to define the nature of Philadelphia society. In such a world, where dependence rather than independence and competency had a purchase on the future, Duane's warnings about the English-like drift of American society seemed especially apposite. Perhaps it was true, as one small craftsman claimed, "that the time before long will come when we shall tread in the steps of Old England."[126]

IV

The economic changes that overtook Philadelphia in the early nineteenth century intensified the divisions that already existed within the city's laboring-class ranks. As Philadelphia industrialized, growing numbers of journeymen experienced the breakdown of a way of life centered on their craft and the small producer tradition. In place of the master who educated his apprentices and guided his journeymen in the "mysteries of the trade," city journeymen increasingly faced employers with little concern for craft customs and eager only for quick and low-cost production. The first decade of the nineteenth century was thus a time of transition when masters and journeymen first sorted out the true meaning of the words employer and employee, a time when the bonds of custom and craft ritual, though weakening, continued to press their obligations on master and journeyman alike.

An early sign of this intra-craft divisiveness appeared in 1801, when the congregation of St. George's Methodist Episcopal Church divided into opposing middle- and laboring-class factions. [127] From the late 1780s, urban evangelists had attracted growing numbers of middling craftsmen to the Methodist cause by appealing to the latent religiosity of Philadelphia's craft community, but also by offering a "respectable" alternative to the traditional laboring-class way of life. [128] By the turn of the nineteenth century, St. George's minister, Lawrence McCombs, controlled church affairs through an alliance with the middle-class membership and a rising group of entrepreneurial masters and small employers. Under his administration, St. George's began to take on a respectable, middle-class mein. This all changed in the spring and summer of 1800 when a citywide revival added more than three hundred ordinary craftsmen to the church rolls. Suddenly McCombs found himself confronted by a group of poorer craftsmen who had been attracted by the fiery rhetoric and visible enthusiasm of streetcorner evangelists and expected nothing less from their permanent church and its hesitant minister.

Anxious to maintain the respectable reputation of his church, McCombs not only rejected the demands of his laboring-class converts but attempted to quell their enthusiasm by removing three of their most popular class leaders. It proved to be precisely the wrong thing to do. Far from having its desired effect, McCombs' authoritarian gesture only increased the distance between his old and new congregants and made the church into the scene of heated and very unchristian disputes. Ultimately miffed by the Methodist Annual Conference, which refused to intervene on his behalf, in June 1801 McCombs led the faction of merchants, retailers, and master-employers out of St. George's Church to establish a new and more respectable congregation at the Philadelphia Academy. Of those who remained at St. George's, most were journeymen and poor independent artisans.[129] The two congregations would never meet together again.

While the issues that brought the dispute about are clouded in a haze of innuendo and self-justification, the cause of the schism itself was clear. Beneath the public rancor and private recriminations, the source of the schism lay in mounting divisions within the craft community itself. The masters and employers who sought redemption and middle-class respectability in early nineteenth-century Methodism simply could not abide a church dominated by the rough voice and enthusiastic displays of their journeymen and employees. The attitude of St. George's departing masters was recorded by John Watson, a "respectable" Methodist and an early chronicler of Philadelphia's secular and religious life. Calling upon the authority of John and Charles Wesley—themselves no friends to enthusiasm— Watson castigated workingmen who in "their simplicity" called upon "the unprofitable emotions of *screaming, hallooing* and *jumping*" to "bring *discredit* on the work of God."[130] These enthusiasts were, he thought, "mostly persons of credulous, *uninformed* minds," men "of rude education" who were "careless of those prescribed forms of good manners and refinement" that were the bedrock of respectability.[131] Such men, he maintained, were no different than "the illiterate *blacks* of the society": they were people "who have the least wisdom to govern themselves" and were thus condemned to "live careless and sometimes trifling lives."[132]

Two years after the schism in St. George's Church, another group of Philadelphia Methodists organized an Hospitable Society in order to proselytize among the city's laboring poor. Their experience, too, reflected the growing rift within the craft community.

In explaining the need for yet another charity in a city well known for its benevolence, the organizers of the Hospitable Society noted that "though many public institutions and private associations, for charitable and benevolent purposes, have already been established in this city, almost every corner presents numerous instances of human wretchedness and misery, spiritual as well as temporal."[133] They went on to note "the distressed situation of the poor in this city [which] few persons are acquainted with, nor are they likely to be, while the suburbs, alleys, lanes, and even some of the streets, in which [the laboring poor] are found, are sufficient by their appearance, to deter many from passing through them."[134] The members of the Hospitable Society were well placed to make these claims, for the officers and visiting committees of the Society were themselves craftsmen, though mostly independent artisans and master craftsmen.[135] As employers and participants in Philadelphia's laboring-class life, the members of the Hospitable Society were closer to the laboring poor than the middle- and upper-class merchants and professionals who maintained the city's other benevolent societies. Thus, while they were concerned as Methodists that the care of the working poor not be left to those secular "persons actuated merely by motives of humanity," they were also aware that "the small pittance allowed by the overseers of the poor is found barely sufficient to purchase bread for a large family."[136]

What the visiting committee found in their rounds through the city's back alleys and suburbs gives us some measure of the conditions in which perhaps as many as a third of Philadelphia's laboring-class families lived. Visitors found journeymen and their families living "in rooms with shattered furniture, in tenements almost in ruins, [some] laying on straw with a few rags to cover them." Others found "more than one family living together . . . in a cold damp abode, with no other furniture than two or three uncomfortable beds, which also served them for chairs and tables."[137] While these may have been descriptions of the living quarters of the worst-placed among Philadelphia's craftsmen, such conditions were real to many and a constant threat to many others.[138] Such conditions, whether real or only threatened, may help to explain the political skepticism of so many Philadelphia craftsmen in these years when neither Federalists nor Democratic-Republicans offered security of employment, reasonable wages, or a guarantee of family subsistence. What might pass for the cynicism and apathy of poor laboring men was in fact the result of the daily struggle for existence reflected in the Hospitable Society's reports. That this struggle for daily subsistence and self-respect tended to focus the attention of many laboring men and their families on the mundane discourse of the stomach rather than the lofty rhetoric of good government is no brief for criticism. Food and shelter of necessity precede the concerns of political life.

Yet whether it was the experience of hunger, inadequate housing, fluctuating employment, and falling wages, or the behavior of middle-class radicals whose attention and interest seemed to draw them more closely to the masters of the city than to journeymen and poor craftsmen, the laboring people of Philadelphia met adversity with a sense of their own dignity and a continued commitment to the community of small producers. It was this closing of ranks in times of distress that in large part explains the hostile reception received by the Methodist visiting committee. For although the committee described its motives in the most caring and benevolent of terms, as coming "with relief to their fellow creatures in [their] wretched abodes," to the journeymen the men of the Hospitable Society were outsiders, a collection of masters and employers who were creating the very problems they had set out to solve. No wonder the members of the Society found themselves greeted with "frequent insults and abuse" as they made their rounds through the poorer sections of the city. [139]

The cold reception given to the masters of the Hospitable Society was a sign of transition and redefinition within the craft community. In receiving hostility rather than the gratitude they expected, the visiting committee discovered in a personal way the depth of the divisions which existed within the city's laboring-class ranks. Shocked by their reception, the visiting masters were nonetheless forced to recognize an important truth: the mutuality of interests and the unity of purpose that had characterized Philadelphia's laboring-class movement a generation earlier had come to an end. In turning away the Methodist masters, the city's poorer craftsmen were serving notice that the paternalism offered by local masters was no more welcome than that proffered by Philadelphia's aristocratic merchant elite. The rough language and impolite gestures with which they delivered their notice hints as well at the growing conflict between the direct and often profane culture of the city's poorer craftsmen and the more polite and circumspect culture that would soon come to be the benchmark of the "respectable" artisan. [140]

There were rumblings of crisis within Philadelphia's craft community at the turn of the century, of which the schism at St. George's and the aggressive reception afforded the Methodist visiting committee give us a first glimpse. The world of the craftsman was being turned inside-out in those years by a process of change that made merchants the employers of skilled craftsmen and master artisans the breakers of custom and craft traditions. In the end, it was a crisis of interest and culture in which we can detect the beginnings of self-definition on the part of members of a working class.

The personal meaning of this crisis is manifest in the testimony given by Job Harrison before Philadelphia's Mayor's Court during the 1806 cordwainer's conspiracy trial.[141] A journeyman shoemaker, Harrison emigrated from England in 1794 and, after plying his trade in Germantown for two years, returned to Philadelphia in 1796 where he took employment with John Bedford, a master shoemaker engaged in the putting-out trade. Every Saturday afternoon Harrison walked from his home to Bedford's shop and exchanged his weekly stint of shoes for wages and more unworked leather. It was soon after he entered Bedford's employ that the tramping committee of the Journeymen Cordwainer's Society first called upon him to join their ranks.[142]

In Harrison's mind the journeymen's society was closer in purpose and structure to a friendly society than a union, the latter being a form of organization altogether familiar to a recent English immigrant.[143] The tramping committee explained that a member of the society "would not work in the same shop, nor board or lodge in the same house, nor would they work at all for the same employer" who hired a non-society man. But Harrison did not view this demand as either unusual or outside the bounds of customary craft relations. As he explained, "I was a man with a large family, and wished to conform to the laws and be a good member." He was, he claimed, "as willing as any one to support the body."[144] When asked to join the society and agree to its purpose and regulations, Harrison expressed neither resentment nor a feeling that his freeborn rights had been transgressed. Rather he subscribed to the old craft idea that the members of a trade ought to organize themselves and regulate the workings of the trade itself.

In this he was following in a long tradition shared widely among the craftsmen of Philadelphia.[145] Even when Harrison fell out with the society over a sympathy strike with which he did not agree, his thoughts and sentiments were those of a member of a craft society. He first "remonstrated with the society at large," arguing that he "had a sick wife and a large young family, and that, I knew I was not able to stand it." When "all the remonstrances I could make were of no use," he decided secretly to become a scab.[146] His reasons for refusing to support the strike, which asked higher wages for boot-makers whose rates were below those of shoemakers, demonstrate the craft context in which he viewed the journeymen's society. He testi-fied that he "did not desire more wages than I then got, more could not be looked for, nor more could not be given."[147] Here is the craftsman drawing upon his knowledge of the trade, the costs of materials, and the market in shoes to make his own, independent judgment of the society's demands. That he found them wanting is less important, however, than the eighteenth-century context of his reasoning. Job Harrison was a craftsman who lived "from hand to mouth, and in debt," and whose "family must perish, or go to the bettering house," unless he continued to work.[148] Rather than face the destruction of his narrow competency, which indeed he thought the society unreasonable for not recognizing, Harrison chose to work outside the bounds of the journeymen's society.

Eventually, Job Harrison was discovered and left the society, al-though again without apparent rancor. When the tramping commit-tee visited his employer, John Bedford, to demand his dismissal, Harrison "expected [Bedford] would knock me off, and I was afraid if he dismissed me that I could not get another seat in the city." But rather than dismissing Harrison, Bedford promised that "we should sink or swim together, if they drive me out of the trade I will turn my shop into a dry good store."[149] Bedford was good to his promise, although he lost several wholesale contracts worth $4000 per annum in the process and faced near-certain ruin for his stand. Nor was John Bedford alone, for the master cordwainers' society also agreed to support journeymen who stayed on with their employers.[150]

In the event the master cordwainer's strategy proved ineffectual and when several masters broke ranks and agreed to meet the journeymen's terms, the rest were forced to follow. When Bedford continued to have difficulty attracting journeymen, he asked Job Harrison to approach the journeymen's society with an offer to pay a fine and thus free his shop from the journeymen's labor boycott.[151] After enduring a brief period of ritual humiliation Harrison paid a small fine and was again accepted into the journeymen's society while Bedford's shop again filled with working journeymen. For his part, Harrison appeared not at all embittered by the experience; despite the humiliation he suffered while arranging his reinstatement, he could claim that "they know I am a well wisher to the body." Moreover, although he did not think the society "had a right to sacrifice the interest of the body" for the sake of bootmakers alone, he maintained that "so friendly am I to the institution that if it was broke down today, I would endeavor to raise it up again tomorrow."[152]

We can detect in these events a craft and its traditions in the process of dissolution. In the testimonies of Job Harrison and John Bedford we are given a unique glimpse into the personal meaning of the process of early industrialization. Harrison and Bedford experienced these events not as the formal subsumption of labor to capital, but as a crisis in their daily lives, a crisis with all of its attendant confusions, contradictions, and insecurities. Through the record of the cordwainer's conspiracy trial we are made privy to an important human drama. There is the antagonist in Bedford, the master craftsman who fought to maintain the traditional relationship between master and journeyman even while he functioned as a putting-out master struggling to meet the competition of other masters and a greatly expanded wholesale market. In Bedford we find the best of master-journeyman paternalism wrapped in the contradictory cloak of early industrial capitalism. There is also the protagonist in Job Harrison, a man who lived by the code of craft custom and tradition at the very moment when that tradition was rapidly being eroded by the pressure of proletarianization. Finally there is the grey figure of George Pullis, leader of the journeymen's society, who in some way understood the logic of the process in which they were all enmeshed and sought to make the journeymen's society over into a class-conscious trade union.[153] We see them here, poised uneasily between a way of life with a long past and one with an uncertain future. It was this mixture of economic roles and personal expectations, of capitalists who wanted to remain traditional masters, of proletarians who thought of themselves as craftsmen rather than simple workers, of far-sighted leaders who sought to translate their prescient vision of the destructive power of industrial capitalism into effective counter-organization, that defines this period of early industrialization in its most human terms. It was this intimate and personal upheaval that called many Philadelphia workingmen away from the political theater in this age of transformation and placed them onto a more prosaic stage. Here they would remain, acting out the subtle shades of craft tradition, until events again called them onto the civic stage.

CHAPTER 6

Confronting Industrialism: Artisan Politics and Philadelphia's Socialist Tradition, 1810–1820

The fissiparous history of Philadelphia's Jeffersonian party in the first decade of the nineteenth century, in which first one and then another element of the popular coalition declared itself the true bearer of democratic values, throws into striking relief the fundamental weakness of the city's radical–laboring-class alliance. Yet even as the city's Democratic-Republicans fragmented into "Quids," "Quadroons," and then "Snyderites," the bonds knit between William Duane's propaganda, Michael Leib's Old School organization, and Philadelphia workingmen continued to hold. Two weeks before the 1810 election Stephen Simpson, son of the chief cashier at Stephen Girard's bank and a young follower of Duane, commented on the political condition of the working class in the Old School stronghold of the Northern Liberties. He conceded that in the preceding decade some workingmen had voted for Federalist candidates but allowed that they were "the honest and well meaning part of [Federalism]," by which he meant "the middling and poorer class of people" who had been drawn in by Federalist rhetoric and tactics.[1] Simpson attributed this laboring-class support for a party "led by men who were notoriously and wickedly opposed to our liberties during the American revolution [and] who aided Lord Howe when in Philadelphia" to "the dependence which it is evident [Federalists have] in a great measure created, especially among many worthy tradesmen and mechanics."[2] By controlling "the chain of connection which exists between the merchant, the mechanic, and tradesman, down to the laboring man," he argued, the Federalists had been able to claim a portion of the laboring-class vote.[3]

But while Federalist influence needed watching, a greater threat came from those apostate Democratic-Republicans who were coming to be known in the second decade of the century as New School Democrats. Led by A. J. Dallas, a Jamaica-born lawyer, one of the founders of the Democratic Society of Pennsylvania, and, since 1791, Secretary of the Commonwealth, these men had been attempting to drive an ethnic wedge between Irish and German workingmen since the 1808 election.[4] Their aim in the coming election, Simpson warned, was to "dislodge democracy from her firmest hold in Pennsylvania" and to replace it with New School rule.[5] Simpson hoped to insure laboring-class unity in the forthcoming election by reminding the workingmen of the Northern Liberties that "the influence of the first congressional district" was well known and that by rallying to Duane's Old School banner, they could bring about "the downfall of men unfit to represent the people [and] secure to democracy honest representatives."[6] Invoking the memory of the English and American revolutions, he reminded Philadelphia workingmen that in the forthcoming election they were contending "for the old democratic cause . . . and for the support of principles produced by the worthies of '75 and '76." Their opponents were, he wrote, nothing less than "an aristocracy of wealth," Tories whose "insidious and unremitted exertions [have allowed them] to realize fortunes at our expense, while they disseminate principles inimical to our rights and liberties."[7]

Simpson's plea to the workingmen of the Northern Liberties was not simple election bombast; the election of 1810 was a crucial test both for the radical–working-class alliance and for Philadelphia democracy as a whole.[8] For more than a decade the Duane-Leib machine had been able to claim the allegiance of small masters and journeymen, who had looked to it as a bulwark against the Federalism and aristocratic designs of the city's wealthy merchants. But the changing structure of opportunity brought about by Jefferson's embargo of 1807–9 placed strains of a new sort on the alliance. As retail merchants began to sell off their stocks of imported cloth, nails, hardware, and other English manufactured goods, some Philadelphia masters sought to refill retailers' shelves with domestic substitutes produced by hired workers. Yet as these prospective employers quickly discovered, the transition from the role of master to that of manufacturer required more capital and credit than most could muster from their own resources. Turning to the city's banks for the necessary capital, these aspiring manufacturers found that banks, even the newly created Mechanics' and Manufacturers' Bank, were partisan institutions that granted loans to a restricted circle of political loyalists and established customers.[9] Thus after 1808 the creation of new banks, accessible to men of moderate means, fast became a dominant and, for the radical–laboring-class alliance, a dangerous issue.

From the creation of the Democratic-Republican coalition in the 1790s tensions had existed between the laboring-class wing led by Duane and Leib and the merchant-Republican wing of George Logan, Tench Coxe, Thomas McKean, and Mathew Carey. Closer in outlook and interest to Federalists than Democrats, the merchant-Republicans had joined the coalition in part because they disagreed with the program of internal taxation that the Federalists had established in 1794, but also because they had been denied access to the center of Federalist power held by the city's merchant magnates.[10] By 1804–5 the tensions that naturally existed between men of substantial wealth—lawyers, shipbuilders, auctioneers, and international merchants—and men struggling for a simple competency brought the merchant-Republicans to repudiate the coalition and to form their own brand of Democratic-Republicanism. This small, breakaway wing, popularly known as the Quids, occupied themselves during the remainder of the decade piecing together pockets of support into what they hoped would become control of the state's Democratic party. With the emergence of the bank issue in 1809–10 they discovered a powerful weapon with which to attack the radical laboring-class wing of the party.

Christening themselves the New School Democrats, leading

Quids organized the Bank of the Northern Liberties in late 1809 to provide credit to the more substantial master craftsmen of the laboring-class district and to any others who hoped to follow the path from craft to manufacturing production.[11] Motivated more by political than purely economic concerns, it was clear from the beginning that the debate over the Bank of the Northern Liberties would involve more than the immediate issue of its charter. For both Old School and New School Democrats, the real issue was the broader and more fundamental question of whether the creation of new chartered banks would benefit or harm the working community itself. Like most other eastern and western states, Pennsylvania had chartered an unprecedented number of banks after 1800, many of which rested on precarious financial foundations. The contest between the Old and New School Democrats over the Bank of the Northern Liberties thus became as much a debate on the probity of the state's banking system as one involving the merits of the Bank itself.

Surveying Pennsylvania's early bank wars in 1816, Mathew Carey, the New School's foremost economic theoretician, summed up the pro-bank position. While there had been irregularities in the operations of some of Pennsylvania's more than forty banks, he admitted, the unsound banks were but "the wild and fantastic projects of a few men, not half a dozen" in number.[12] Chartered banks were, he assured his readers, sound and beneficial institutions. Although improperly run banks could produce and foster "luxury, extravagance, and speculation," without them, he predicted, "the middle and lower classes of society" would face ruin.[13] For Carey and Philadelphia's New School Democrats, proper regulation of the banks would guarantee their usefulness to the entire community and prevent their turning into engines of exorbitant accumulation. The question was thus not one of number, but of the regulation of state banking. As Carey emphasized, "there is one important view of the relations between the community and the banks, of which these institutions should not for a moment lose sight. They are as entirely at the mercy of the public, as the public are at theirs: and they hold their standing by the tenure of general indulgence."[14]

If the New School held what historian Louis Hartz has labeled pro-charter views, the Old School did not fit as easily into his depiction of the anti-charter position.[15] For Hartz, the anti-charter position was characterized by "a puzzling aspect of unreality" caused by the intellectual distance between the "ideology" of anti-charter advocates and the practical values of the public at large.[16] According to Hartz, anti-charter proponents were ideologues who opposed banks because they viewed them as undemocratic instruments which raised the economic standing of one portion of the community to the disadvantage of the community as a whole. Against this view, Hartz claimed that public opinion "bravely defended" charters because their very impersonality removed them from "the pettiness and conniving" of individual banks and bankers.[17] Thus, in the Hartzian view, the ideology of the anti-charter advocates ran against the more pragmatic outlook of the larger public, whose faith in the "competitive principle" contrasted with the more ideological views of their leaders.[18]

The reality of Philadelphia political life contrasted sharply with this view. William Duane and Stephen Simpson, the Old School leaders closest to Hartz's anti-charter position, were not the backward-looking and ineffectual ideologues he portrays. They were, in the first three decades of the nineteenth century, powerful political leaders with a substantial laboring-class following. Moreover, the anti-charterism of Old School Democrats only developed during the bank mania of the second decade of the nineteenth century, as Duane, and later Simpson, moved from a pro-bank position similar to Carey's to take a hard-money and anti-bank stance.[19]

The most mature statement of the Old School's banking position came from Stephen Simpson, who argued in his *Workingmen's Manual* that the state's banking system was founded on simple fraud. A bank bill, he wrote, is merely "an order on paper, for so much money, *drawn upon the producer of labour* [and] paid by public credulity, faith, or what is sometimes called credit."[20] "The party that profits, and the only one," he continued, "in this transaction of fiction and fraud, is the banker, the stockholder, and the speculator." Through the banking system, Simpson claimed, "banks, or corporations, make use of the entire property of the community for their own exclusive profit, interest, and usury." In the end, "it is *labour* that pays the bankbill, it is labour that pays the interest," and it is labour that suffers "under the double burden of sustaining the idle, pampering the rapacious, and gratifying the gambler."[21] In its most developed form, then, the Old School viewed the contemporary banking system as yet another means by which labor was extracted for the support of non-producers. Because of this, banks required more than supervision and regulation; they needed to be founded upon a different and non-exploitative system. This is a large distance from Hartz's view of the anti-charterists as idiosyncratic cranks who habitually set their faces against any form of progress. Duane and Simpson opposed the creation of new banks not because of their antipathy to things modern, but because the banking system of the early nineteenth century exploited Pennsylvania's producing classes. The public opinion which Hartz found overwhelmingly supportive of charters and corporations was in reality the opinion of New School merchants and manufacturers, and not that of the ordinary workingmen who rallied around the Old School banner for nearly thirty years.

Duane, for his part, wrote forcefully against the general proliferation of banks that accompanied Jefferson's embargo. He pointed to the financial dangers that severely undercapitalized banks held for their depositors as well as for the nation itself. More important, he argued that those who had the most to gain from the bank mania were not middling and lower-class producers but speculators and the idle rich.[22] Always attuned to his own constituency, Duane reminded the city's workingmen that a large number of small banks, each emitting its own currency, would inevitably lead to a depreciated currency and lower real wages for journeymen and wage-earners alike.[23]

Thus the scene was set for the political counterpart of the economic division between masters and journeymen that had occupied the previous generation of Philadelphia craftsmen. Much as the diverging interests of masters and craft workers had been expressed in the creation of separate trade societies, the 1810 election was poised to test these interests in the political arena. Through the creation of the Bank of the Northern Liberties the New School moved to divide Democratic craftsmen along master-journeyman lines and to claim the master-manufacturers for themselves.

Writing on the eve of the election, "Marcellus" put the matter plainly. "The democratic party is on the eve of dissolution," he told the city's workingmen, "and by your union and energy alone can it be saved."[24] Laboring-class unity was, however, conspicuously absent on election day, as Philadelphia workingmen split their votes between the New and Old School tickets and allowed the Federalists to dominate the city election for the first time in over a decade. The division within laboring-class ranks was especially pronounced in the bulwark of Old School power, the Northern Liberties. There the New School congressional candidate, Dr. John Pinton, received 61 percent of the vote compared with the Old School's meager showing of only 39 percent.[25] Even worse, in Philadelphia County, where the bulk of the city's workingmen lived, New School candidates amassed 43 percent of the vote while the long-dominant Old School polled a disappointing 34 percent, barely 11 percent more than the Federalist ticket.[26]

For Duane, who understood city craftsmen better than any other democratic leader, the election results caused little surprise. Surveying the new political situation a few days after the election, he claimed that he had "observed silently the operations that were passing and pretty accurately foresaw the result."[27] The unusual silence of the *Aurora* during the 1810 campaign was an indication that Duane and the leaders of the Old School alliance understood the cleaving pressures that worked to undermine the bonds of their coalition, just as Simpson's lengthy appeal to the Old School's traditional Northern Liberties constituency was an admission of the tensions dividing the city's most loyal workingmen. The fact that laboring-class Federalists were treated with such compassion and understanding by Simpson, especially compared with the spleen he vented on the New School leadership, served only to underscore the depth of the division. Never before had an Old School leader had to plead for political unity and laboring-class support.

The New School victory in the Old School stronghold of the Northern Liberties marked both an end and a beginning. The fracturing of the laboring-class vote and the growing dominance of the cash nexus in master-journeymen relations that it, in large part, represented announced the demise of Philadelphia's laboring class in its eighteenth-century form. Before 1810 the distinction between master and journeymen had been less important in understanding laboring-class life than differences between trades, ethnic origins, or religious affiliations. In the eighteenth century these internal divisions had been overridden by an enduring sense of mutuality and common social position which found expression in craft traditions, public rituals, and the moral tenets of the small-producer tradition. By the second decade of the nineteenth century, however, Philadelphia craftsmen shared neither a common culture nor similar interests; industrial capitalism was beginning to leave an indelible mark on social relations in the Quaker City.

The nineteenth century brought new economic opportunities and a rising class of capitalist manufacturers to exploit them. The pressure of enlarged markets and punishing competition worked powerfully to erode the old reciprocal relationships between masters and journeymen that had promised good workmanship in return for ample employment and fair pay. Employing masters now found themselves pressed to reduce wages, prolong working hours, hire the cheaper labor of unapprenticed and partially trained men, and subdivide tasks in their shops. The grinding stone of competition worked especially hard on those whose relation to "the trade" was always marginal and whose continuance in it was always tenuous. Outwork and garret masters racked the rents of their looms and wheels, found imaginary "faults" in their worker's weekly returns in order to lower their piece rates, and in general found the most intricate and petty reasons to cut their costs and increase the production of those in their employ.[28] This was the new world prefigured in the 1810 election. It was the world that William Duane had warned against three years before in his "Politics for Farmers and Mechanics": the English working-class experience recreated on American shores.

The little-noticed election of 1810 thus possessed a double importance. First by drawing a sharp line between the merchant-manufacturing and the traditional craft wings of the Democratic party the election made a unique contribution to the formation of Philadelphia's working class. The tension that had been building within the city's laboring ranks now took on a plainer and more portentous meaning. In the New School branch of the Democratic-Republican party journeymen could now recognize those against whom they were struggling in the workplace as their political foes as well. Rather than being the "coffins of class-consciousness," as one recent historian has claimed, Philadelphia's party politics added another dimension to the class experience and accelerated the development of a distinctive working-class consciousness. [29] In light of much current historiography it is necessary to stress this point. In Philadelphia, if not elsewhere, local political contests raised incipient economic divisions, hitherto isolated in the experiences of craftsmen within their individual trades, to the threshold of class consciousness. In the same way that the betrayal of working-class claims by the reformed parliament of 1832 projected the class structure of England in high relief and ignited the Chartist movement, so the betrayal of journeymen's interests and their small producer values by New School Democrats in the election of 1810 highlighted Philadelphia's changing class structure and opened the path toward the workingmen's movement of the following decade. [30]

The election of 1810 was important in a second way as well. The division of Philadelphia's Democratic party into Old and New School wings reveals at an early stage of development the peculiar nature of American industrialization and working-class formation. For the rivalry between Old and New School Democracy was more than a struggle over offices and political spoils; it was also a contest over the future contours of American society. Both the Old and the New School wings agreed about the need for the development of domestic manufacturing; what they disagreed about was the role that workingmen were to play in the process of industrial development. [31] For the New School, the key to industrial development was the manufacturer himself, the man who supplied the fixed capital of workplace and machinery that would allow American industry to compete with British manufactured goods. In one of his many discourses on the subject, Mathew Carey sang the praises of the domestic manufacturer in typical New School fashion. "Their operations were commenced under great difficulties," he wrote, "and [with] various discouraging disadvantages of the most serious character." Not only did the owners of mill sites sell them "at most extravagant prices," but mill wrights and experienced workmen "could not be had but at very high wages." In their struggle to establish manufactories, Carey concluded, "almost every person in the nation with whom the manufacturers had to deal, took advantage of their necessities, and made them victims of extortion."[32] In Carey's hands, the manufacturing bias of the New School is manifest. Workingmen and women enter the New School equation as unprincipled extortionists, forcing besieged manufacturers to pay more for their labor than it is worth. And, in a radical inversion of the labor theory of value, self-serving workers use the scarcity of skilled labor to exploit the nation's struggling and beleaguered manufacturers.[33]

The Old School viewed manufacturing from a different vantage point. Their focus was not on the manufacturer, but on the common craftsman whose labor was the source of all national wealth. For the Old School, machines and manufactories were the products of previous labor and added nothing to the common stock of value. "Labour saving machinery," Stephen Simpson argued, was "compound labour," the embodiment of past labor which greatly multiplied the productivity of living labor. "Machines," he wrote in 1831, "require no clothing, no food, [and] attended with little or no expense, produce a hundred, or five hundred fold the profit of human labour." That the most mechanized nations were also "the most opulent and powerful," only confirmed the fact that "LABOUR is the source of all national greatness, as well as individual happiness."[34] Given this emphasis on the labor of working people, it is not surprising that the Old School program for manufacturing development called for a community of producers who would employ machinery not only to increase production but to diminish the time and burdens of labor. "Whatever saves time and labour," Simpson wrote, "must by so much increase the sum of our comforts and property, and add to the general happiness. All improvements in the *instruments* made use of by mechanics . . . which lessen and abridge labour, add to the stock of industry and extend the general enjoyment." Machinery, Simpson declared, held forth the prospect of a new age for labor. "If we can produce twice the labour in half the usual time, we become richer, and have leisure to cultivate our intellect and enjoy the beauties of nature." "It is by compound labour," he predicted, "that the working classes are destined . . . to rise in the scale of science, intellect, and knowledge."[35] Reduced to its essentials, the distance between the Old and New School views was the infinite space that separated exploitative and communitarian visions of manufacturing development.

The Old School and New School visions of industrial development reveal the great peculiarity of American working-class development. In a curious inversion that would characterize the American labor movement to the eve of the twentieth century, the New School represented capitalist industrialization cloaked in eighteenth-century paternalist-deferential garb while the Old School came increasingly to represent pre-industrial small-producer traditions in the costume of class-consciousness. It was this tension of old and new social relationships which ran through both wings of the Democratic party and stamped the character of Philadelphia's working-class movement. Unlike Britain where an ancient landed aristocracy could become the repository and standard-bearer of traditional social relations, leaving the emerging class of industrial capitalists free to follow purely economic ends under the banner of Malthus and Bentham, in America the ideals and affective bonds of the past were a powerful and enduring force which confronted capitalist and worker alike.[36]

II

Some notion of the depth which divided the journeymen and small craftsmen of the Old School and New School merchants and manufacturers can be gleaned from the Democrat's response to the second Anglo-American war. The declaration of war on June 18, 1812, gave a boost to Philadelphia's Federalists who saw in wartime patriotism the means to secure the tenuous hold they had gained over city politics in the 1810 election. But while the Old and New schools were at loggerheads over their approaches to industrialization and differed in their sentiments about the ordinary workingman, they were nonetheless Democrats and the heirs of urban Jeffersonianism. Accordingly in the 1812 elections the Old and New schools held their differences in check and united in opposing the Federalist ticket. The result was a sweeping victory that placed Democrats once again in overwhelming control of city and county offices. [37]

But though wartime unity had produced a dramatic victory, patriotism proved to be a poor cement with which to bind the diverging parties. Even as the rival Democratic factions met together in September 1812 to hammer out a united ticket, the Old School, under the direction of Michael Leib, bolted from the unity ticket and replaced four New School candidates with four of their own. [38] It was the first sign that prewar tensions could not be contained by the exigencies of war.

At first glance things appeared calm among the city's workingmen. Early in 1813 William Duane, now Adjutant General of the state militia, called for volunteers to man the city's ramshackle defenses and 4000 workingmen quickly responded, each pledging a day's work without pay in defense of their city. [39] The sheer size of the laboring-class response indicated that this was no partisan undertaking, and the work crews were made up of Old School and New School adherents as well as laboring-class Federalists.

One explanation for the relative calm of the early war years was no doubt the economic boom produced by the war itself. Much like the colonial and Revolutionary wars before them, the second Anglo-American war offered good times to Philadelphia workmen. Wartime demand increased the need for labor, and employers offered full employment at ample wages to a large majority of the city's craftsmen. [40] At the same time the absence of British imports spurred the development of domestic manufacturing in the city and county as weavers and metalworkers joined other craftsmen in providing the cloth, uniforms, nails, and shot that no longer flowed from English factories. [41]

War contracts, too, created new and enlarged fortunes for military suppliers, just as they had in the eighteenth-century wars. As one observer accurately noted, "the whole host of army contractors, &c., drink in full bumpers [to the] 'duration of the war.' It is to them *meat, drink, washing, and lodging.*"[42] But while profiteering was a common wartime pattern, the laboring-class response was unusually muted by Philadelphia standards. In fact, the workingmen's only concerted response came in late 1813 as the prices of consumables such as coffee, sugar, and salt rose beyond the means of laboring-class budgets. In response Philadelphia craftsmen formed consumer associations, modeled on the popular committees of the Revolutionary era, whose members pledged not to purchase goods above their prewar levels.[43] But while these associations again raised the prospect of conflict between the city's merchants and laboring-class consumers, price controls proved superfluous, as talk of peace spread throughout the city in the spring of 1814 and merchants and retailers lowered their prices in anticipation of renewed trade with England.

Better economic circumstances thus touched nearly everyone in wartime Philadelphia. Master mechanics and manufacturers saw in the rise of domestic industry and the proliferation of banks the means both to create and to finance their enterprises. Small craftsmen and wage-earners saw in the war new opportunities to better their position and, in the case of some journeymen, perhaps a chance to escape the dependency of wage labor. In the last analysis, the War of 1812 appeared to be Philadelphia's least disruptive war.

In fact only two groups suffered tangibly during these boom years. The first was the families of soldiers and sailors who depended upon the vagaries of military pay for their support. The saga of the American military in the second Anglo-American war was an unremitting story of incompetence, disorganization, and mishandling.[44] Called away from their families for militia or national service, common soldiers found that their hitches were filled with a great deal more disease and boredom than either the thrill or terror of combat.[45] Compounding this, soldiers also discovered that their pay came at maddeningly long and uncertain intervals, leaving their families unprovided for and forced to rely on the city's public and private charities for simple support.[46]

Faced with a growing number of poor families and a continuing lack of federal support, by 1814 the Philadelphia Committee of Defense, a body charged with running the city's military affairs, was forced to undertake a private subscription to relieve the palpable distress of the soldiers' and sailors' impoverished families.[47] Even then, despite Philadelphia's reputation as a benevolent city, relief subscriptions trickled in at such a slow rate that by year's end the Committee had less than $5000 to distribute to poor soldiers' families.[48] In the end, it was left to the city's Democratic ward organizations to care for the families of Philadelphia's soldiers and sailors and to call upon the wealthy of the city to perform their civic duties.[49] As Duane admonished Philadelphia's "better sort," "those whose houses are to be defended by men in humble life—those who have wealth—should aid the wives and children of such as are in want of it."[50] It seemed to be the only simple justice.

The second group that failed to partake of wartime prosperity was Philadelphia's emerging community of free black workers. Confined to the most menial of occupations—laborers, mariners, carters, and sawers of firewood—the benefits of the heated wartime economy touched them less than it did their white counterparts. The tangible poverty of laboring-class Blacks led "A Volunteer" to advance, in early 1813, a remedy that at once reveals the difference in black and white fortunes during the war and the unexamined racism of even the advocates of democracy in the early Republic. Writing in the *Democratic Press*, a New School newspaper founded in 1807 by the radical Irish expatriate John Binns, "Volunteer" wrote of the "difficulty in obtaining by enlistment the requisite number of men to carry out the war with vigor and energy."[51] The solution to declining enlistments, he suggested, lay among the city's free blacks, that "large portion of the male population" who contribute nothing "at all to the maintenance or support of government." The "free people of color," he argued, "ought to be brought forward to fill the ranks of the army" in the place of laboring-class whites. Black workingmen were, after all, a "very numerous and useless" surplus population which "could be better spared than any other class of the population." In fact, "Volunteer" calculated, "Pennsylvania alone could well spare two thousand of these people and be benefitted by it more especially as it might leave many farmers and mechanics at their useful vocations, and with their families."[52] That free black men, too, might have to leave their "useful vocations" and "their families" appears to have escaped "Volunteer"'s notice.[53]

If the three years of the second Anglo-American war had relatively little impact on Philadelphia's working people, the same cannot be said of the years from the end of the war to the close of the decade. The year 1815 not only marked the end of Anglo-American hostilities but signaled the close of the Napoleonic Wars as well. Philadelphians whose memories stretched back to the early 1780s might have testified that history does indeed repeat itself. Following the Treaty of Ghent, British goods flooded into Philadelphia and other American markets just as they had in the aftermath of the Revolutionary war. Despite the rapid development of Philadelphia manufacturing during the embargo and war years, by 1816 local manufacturers found it difficult to compete with these cheap imported goods, especially when their quality often exceeded that of domestic products. Writing in 1819, at the depth of the postwar depression, John Hinshellwood, a small cotton cloth manufacturer, testified that because of British competition "markets are so irregular and falling everyday" that if "the markets continue to depress, I will be under the necessity to drop [the business] altogether." As it was, Hinshellwood complained, "we have no other market but Auction stores, as merchants will not purchase privately" from domestic manufacturers.[54] In the same year, James Crosby, who styled himself "A Domestic Manufacturer," described the plight of others of his class. "Owing to foreign importations," he wrote, "we are compelled to sell at cost." "All I make," he lamented, "is by boarding my workmen."[55]

Others would concur. From Boston to Baltimore the postwar inundation of British goods pushed American manufacturers to the margins of solvency. No longer able to rely on the temporary trade barriers created by the embargo and war, masters and manufacturers, undercapitalized even in the best of times, laid off workers at alarming rates in desperate attempts to keep themselves afloat.[56] John Hinshellwood was far from atypical when he reported that he "formerly employed 14 looms [and] now employed 7 only."[57]

By early 1818, some 16,000 metal workers and laborers were without work in Philadelphia County.[58] Shipbuilders fared even worse: in 1816 Philadelphia shipyards produced but one ship, and the following year they produced nothing at all; 1818 proved to be little different.[59] As one observer remarked late in the decade, "the merchants whose exclusive occupation is that of shipbuilding have been languishing for several years past and but for the public service in the Navy Yard, the best workmen in that art would have been destitute."[60] In 1816 a correspondent to the *Aurora* summarized the position of producers throughout the Philadelphia region. "Few, if any of these useful manufacturers who continue to work," he wrote, "have half the numbers of hands engaged which they had only six months ago."[61] Even local farmers, who relied on outwork, by-employment, and the sale of their surplus crops for cash income, felt the pinch of the depression. Not only were they "no longer called upon for that supply of provisions" that had existed before 1815, the correspondent explained, but "their families [could no] longer obtain work, either in their own houses, or at the factories."[62]

Had history truly repeated itself, the post-1815 economic crisis would have resolved itself in much the same fashion as the crisis of the 1780s. Then, demand for domestic production resumed once the slack of British production had been taken up by local consumption and new markets for local goods were found. But the end of the Napoleonic Wars did not bring an increase in foreign demand for American products nor did the pressure of British production abate. The enormous costs of protracted warfare had cost the British exchequer dearly, and tax increases throughout the postwar years drove English manufacturers to reduce costs and increase production to new levels in order to meet their mounting assessments.[63]

To make matters worse, for more than a decade Philadelphia had been the center of a dangerously expanding banking system characterized by a large number of small, undercapitalized, and overextended institutions. In 1814 alone 42 banks received charters from the Pennsylvania legislature.[64] The bank mania of the early nineteenth century alarmed even such a staunch supporter of new banks as William Duane. Before 1810 he had seen small banks as "democratic" institutions which made credit available to small craftsmen and farmers when the larger, merchant-controlled banks would not. But the explosion of speculation that followed the reckless granting of bank charters coupled with the increasingly unsound policies pursued by the new banks brought Duane to a hard-money, anti-bank position by the end of the war.[65]

By 1819 the worst of Duane's fears were realized. Already racked by four years of deep depression, Philadelphians now faced the collapse of the banking and credit systems in the Panic of 1819.[66] Without access to credit independent craftsmen could not purchase raw materials nor could small employers hire workmen. Coming at a time of chronically high unemployment, the collapse of the financial system drove growing numbers of laboring men from their workbenches onto the city's poor relief rolls. In January "A Citizen" complained of the mounting poor taxes *"under which our city now groans"* and the "hundreds who now apply for admittance" to the city's almshouse.[67] By August laborers' wages had fallen so low that even the New School *Democratic Press* called for a "regulation of wages" to insure subsistence for the city's poorest workers.[68]

At year's end conditions were so bad throughout the nation that state legislatures took the unprecedented step of forming public committees to investigate the effects of the depression and to propose ways of alleviating the mounting distress. Faced with "a scene of distress and pecuniary embarrassments unexampled in former times," the Pennsylvania house of representatives appointed William Duane, who had been elected to the assembly in 1818, to chair the state's investigative committee.[69] As expected, Duane's report revealed an economy in acute disarray. "Distress is general," he reported, and "it extends to the former capitalist, as well as to the most humble farmer and mechanic."[70] Among the depression's worst effects, Duane singled out the "destruction of capital, the emigration of our citizens to the wilderness, the stagnation of business, the deterioration of landed property, and the prostration of manufactories."[71] Quoting passages from petitions sent from every corner of the state, Duane underscored the dire straits in which many Pennsylvanians found themselves. A petition from Northumberland County lamented that "the greater part of the citizens, even with the utmost economy and industry are scarcely able to obtain sufficient articles to sustain life."[72] An angrier petition from Huntingdon County declared that "the industrious are impoverished whilst the speculating part of the community are growing daily more wealthy."[73] For most Pennsylvanians, the committee's report revealed, it was the worst of times.

Duane's committee was not alone in investigating the social consequences of the depression. In Philadelphia, the Pennsylvania Society for Promoting Public Economy sent a circular letter of enquiry to the Guardians of the Poor asking them to survey the condition of the laboring classes in each ward of the city.[74] Composed of government officials, clergymen, and prominent merchants who continued to feel a paternalistic responsibility for the poor, the Society sought ways to maintain the "needy" poor while at the same time lowering the poor rates that were making increasing claims upon their pocketbooks.[75] The *Report of the Library Committee*, issued by the Society in 1817, found the city's free blacks, recent Irish immigrants, and day laborers to be the most impoverished groups in Philadelphia. The vast majority of the poor, they also reported, blamed their poverty on lack of employment, to which the Committee added the high prices of fuel and provisions and—ubiquitous among reformers of the time—the intemperance of the impoverished.[76]

The Committee found the plight of women to be even worse than that of men, owing, they thought, to the enormous number of women and families deserted by "enlistees in the late war." There were, moreover, between 7,000 and 10,000 unemployed women and children in the city as a result of the closing of local manufactories. As always, they found the families of sailors to be particularly impoverished. Although the burdens of the postwar depression fell upon nearly everyone, the Committee concluded, the construction and maritime trades—ship carpenters, riggers, caulkers, sailmakers, carpenters, plasterers, bricklayers, day laborers, and boatmen—fared the worst of all.[77]

The gloomy picture that emerged from the Committee's *Report* was confirmed by the returns of the sheriff and prothonotary of Philadelphia County in 1819.[78] Comparing the postwar conditions of working Philadelphians with their situation at the end of Jefferson's embargo, the returns revealed that court actions for debt had increased by 62 percent between 1809 and 1819 and that sheriff's sales of foreclosed property had increased 153 percent in the same period. Imprisonment for debt, one of the most fearsome of laboring-class calamities, rose by over 38 percent in the course of the decade while confessed judgments against debtors skyrocketed from 443 to 1,158.[79]

The accumulated evidence confirms contemporary opinion that the five years following the end of the war was a period of major disruption in the lives of most Americans, but especially those who labored for their livelihoods. As one of Duane's petitioners aptly claimed, "at no time since the revolution has a greater distress been felt than at the present moment."[80] It was the hope of all that they would never see such times again.

III

The close of the second Anglo-American war brought more than economic disruption to the lives of laboring-class Philadelphians. The coming of peace also marked the end of the grudging wartime rapprochement between Old and New School Democrats and revived their contest for political dominance begun during the embargo years. In Philadelphia, the resumption of political conflict revolved around the efforts of both factions to win the city's laboring-class vote. For if the contest for the allegiance of merchants, manufacturers, and employing masters had largely been decided by 1810, the largest section of the city's laboring class—the small, independent masters, the journeymen, and the common workingmen—had yet to be broken from their Old School moorings.

It was with this stalwart constituency in mind that Michael Leib mounted his final campaign to revive the prewar coalition of artisans and journeymen in the Northern Liberties.[81] Drawing upon old loyalties in the laboring-class district that he had served for nearly twenty years, Leib easily won two terms in the state assembly. But despite his immediate electoral success, Leib's long-term inability to deliver the legislation his constituency demanded—in particular, a mechanics' lien law and the abolition of imprisonment for debt—gradually eroded his popular support. Gravely ill, in 1818 he made a last, desperate attempt to augment his dwindling base among district workingmen by allying with a small contingent of moderate Federalists in the Northern Liberties. It was truly a gesture of futility. In the end, Leib's "Amalgamation" ticket was a personal and political disaster as his old supporters joined with younger workingmen from the Northern Liberties to vote against what they saw as a perfidious compromise with aristocracy.[82] With the 1818 election the political machine that had represented the interests of laboring men and had given a public voice to their needs and aspirations for more than a generation had come to an end. Left with few alternatives many Philadelphia workingmen turned to the New School by default, or abstained from politics altogether.[83]

The 1818 election confirmed what the election of 1810 and the widespread existence of independent journeymen's trade societies had already revealed: economic development had proceeded beyond the point where the interests of employing masters and wage-earning journeymen could be encompassed by the mutuality and craft traditions of the past. Jefferson's embargo and the recent war had only accelerated this division by catapulting growing numbers of merchants and master craftsmen into the ranks of the city's industrial capitalists. Under these conditions the eighteenth-century notion of "the people"—small producers, shopkeepers, and laborers united by the local nature and small scale of craft production—was as anachronistic as the Old School coalition that had represented their interests.

The Old School connection with the city's laboring classes did not die in the debacle of 1818, but it did take on new forms and aims. Already in his "Farmers and Mechanics" essays of 1807, William Duane had begun to take on the role of a social commentator and critic. After Leib's defeat in the 1818 election the Old School itself ceased to operate as a political party and became, in effect, a protest movement. Set against the background of an emerging industrial class system, Old School loyalists, many of them younger journeymen from the Northern Liberties and Southwark, fell in behind the intellectual and moral leadership of Duane, his son William John, and Stephen Simpson, a family friend and political protégé. [84] By one count controlling no more than 11 percent of the Democratic votes in the state, the trio set about uniting the remnants of Old School support among the city's workingmen in order to rescue the small-producer world from impending doom. [85] In an appeal that recalled the recent experiences of craftsmen who continued to hold small-producer values, Duane castigated those who had become "poisoned by their own vanity" and had profited from other people's misery. [86] The "disinterestedness" and sense of community spirit that had characterized earlier days, he wrote without exaggeration, "are surrounded by mists and darkness."[87] It was once again necessary to reassert those small-producer values and the way of life they represented.

In the course of the next decade Simpson and the senior Duane hammered repeatedly on this theme. Only a true return to small-producer values and the spirit and promise of the "principles of 1776 & 1800" would restore craftsmen, journeymen, and wage-earners to their rightful places in society. [88] Yet at the same time that they attempted to reawaken the old small-producer ethic, Duane and Simpson inevitably changed it. As central figures in Philadelphia's emerging socialist movement, the two critics helped city craftsmen transform their small-producer traditions from a moral critique of early industrial capitalism into a program for action.

Socialism had always been a logical, if unexplored, corollary to artisan notions of the social utility of labor. If, as the small-producer tradition maintained, labor alone produced wealth, it followed that all social wealth should by right belong to those who labored to produce it. The first glimmerings of this interpretation of small-producer values appeared in the price-control controversy of 1779 when the Committee of Thirteen introduced a notion of "service" that not only questioned the sanctity of private property rights but set the stage for later calls for communal ownership of goods by the associated producers of the city. [89]

Yet, pregnant as this early extension of the small-producer tradition was, the real beginning of Philadelphia socialism came in 1802 with the publication of *Equality: A Political Romance*. Published serially in the *Temple of Reason*, a deist newspaper directed at the nation's "middling and lower classes," *Equality* was the work of Dr. James Reynolds, a Dublin physician who had immigrated to Philadelphia in 1794 after serving a prison term for his participation in the United Irish movement.[90] With credentials like these, Reynolds quickly moved into the inner circle of the city's Democratic-Republican party where he became a close friend of William Duane.[91] The resulting political and personal bond was so strong that in 1799 city Federalists singled out Reynolds and Duane for their first prosecution under the new federal Sedition Act.[92]

The first socialist tract written in America, *Equality* was the literary fulfillment of the small-producer tradition. A fictional account of an American's travels in the utopian nation of Lithconia, it was, in fact, the world of early industrial Philadelphia turned upside down. Unlike the increasingly callous relationships that dominated the manufacturing city, Lithconia represented an ideal community of small producers united in the bonds of true human affection. Reynolds emphasized this contrast in his preface, where he argued that while "Happiness is what we are always seeking after, where industry is wound up to oppression, Society can never be on a good construction."[93] Reynolds's alternative to Philadelphia's faulty social construction was Lithconia, a country where money did not exist, where land was held in common, and where daily labor was required of every citizen. In contrast to the Quaker City, where a ten- to twelve-hour workday was common, Lithconians labored but four hours a day, spending their remaining hours in rest, recreation, and study. Most important, Lithconians lived up to the small-producer ideal, for not only did everyone labor, but Lithconians were prohibited from performing the labor of others as well. Unlike contemporary Philadelphia, there was no wage labor in Lithconia and no parasitic class of non-producers to live on the fruits of another's labor. Lithconia had, in fact, long since rid itself of all forms of unproductive labor by creating community warehouses for the storage and distribution of food, clothing, and household essentials. By distributing these "goods" among Lithconians on a regular schedule, the time-wasting necessity of shopping was eliminated and residents were able to devote even more of their time to learning and community pursuits. Such a system of distribution, Reynolds argued, also relieved society of the need for unproductive members such as merchants and "the shopkeepers which seem so necessary in barbarous countries."[94] In Lithconia, he wrote, "two men distribute as much provisions as half the hucksters, grocers, bakers, and butchers in Philadelphia, and two more can distribute as much clothing in one month as all the Quakers in Pennsylvania will sell in a year."[95]

Reynolds's Lithconia thus offered an image of an ideal small-producer society; a social system that required neither merchant nor middleman and a system of work which did not drain a person's energies through long hours of labor. Added to this, Reynolds held out the prospect of marriage and family life freed from the dominance of man over woman and devoid of the separate spheres in which contemporary men and women carried on their daily lives. Political and religious freedom were, of course, guaranteed in Lithconia, although there was no place for an established church or regular clergy. Even military service was democratic, for unlike the experience of Revolutionary Philadelphia, no Lithconian could avoid service by paying a fine or hiring a substitute.

Although sometimes crudely drawn, Reynolds's Lithconia nonetheless offered to laboring-class Philadelphians a model of an alternative society closely attuned to their needs and devoid of the unemployment, insecure subsistence, and threatened competency that made up their everyday lives. While it is impossible to measure directly the reception of *Equality* among the city's working people, Reynolds's friendship with Duane, his active participation in Old School politics, and his medical practice in the city's laboring-class wards suggest that his ideas received at least a careful hearing.[96]

William Duane himself made a small but critical contribution to Philadelphia socialism in his 1807 essays entitled "Politics for Farmers and Mechanics."[97] Duane had been reading the social commentaries of William Cobbett and absorbing the thoughts of the radical critics of English political economy since the turn of the century. Now as he turned his eye to economic and political developments in Jeffersonian America, he utilized these ideas to move beyond small-producer notions of the dignity of labor and the worth of the workingman to advance the question of exploitation itself. It was a conceptual turning point that would lead to a radical revision of the small-producer tradition and dominate laboring-class thought for the next two decades. Following Paine and Cobbett, who had placed exorbitant taxes and an elite-dominated state at the center of their economic analysis, Duane located the mechanism of exploitation in the developing American taxation system. Through their dominance of state and national politics, Duane argued, merchants, employers, and speculators had created a taxation system that drained surplus labor from producers' pockets and quietly deposited it into their own bank accounts. The result, he pointed out, was that "every *idle* fellow expects to live on the sweat, and the labour, and the industry, and the talents of the virtuous part of the nation."[98]

While ultimately inadequate as a theory of exploitation, by focusing on the extraction of surplus labor from society's producers, Duane nonetheless made a significant advance on earlier explanations of the declining craft economy. Where writers from Benjamin Franklin to the Revolutionary Committee of Thirteen had assumed that non-producers were entitled to a portion of the social product because they had put forward their capital or had served some socially useful purpose, in Duane's essays farmers and mechanics are seen as the *sole* producers of wealth and the income of non-producers is condemned as an unjustified tax on their labor. It was a subtle but important difference. In the context of early industrial Philadelphia, the older view implied that mechanics were worthy of their hire and justified in claiming good wages for their labor, but little else. In Duane's view, on the other hand, producers were entitled to the *whole* product of their labor and anything less was unjust and galling exploitation. While still a crude and undeveloped idea in Duane's hands, the concept of exploitation would mature dramatically in the coming years.

In the decade after 1807 Philadelphia socialism took a different turn in the form of William Maclure's educational experiments. Best known as Robert Owen's partner in the New Harmony experiment of

1825, Maclure was a Scottish merchant who retired from commerce and emigrated to Philadelphia in the early years of the new Republic.[99] Once in Philadelphia, Maclure took up scientific pursuits, performing the nation's first geological survey and helping to found the Philadelphia Academy of Natural Sciences. Maclure was also interested in reform and in 1805 traveled to France where he met with Joseph Neff, a Parisian proponent of Pestalozzian educational reform. Impressed with what he saw, Maclure brought Neff to Philadelphia and installed him as head of a Pestalozzian academy at the Falls of Schuylkill in 1809. At Neff's radical academy students learned by observation and the Socratic method rather than the rote memorization and mass recitation that were the characteristic methods of American education. Most important, Maclure and Neff viewed the school and the Pestalozzian system itself as a means to bring about the radical restructuring of society. "Knowledge is power," Maclure wrote, "and it is . . . impossible to keep a well informed people in slavery." For Maclure the Pestalozzian system would produce free adults ready to use the practical knowledge they gained from their education to create an "equal division of property" and thus give true "vigor to the great mass" of the people.[100] William Duane apparently agreed, for he enrolled both his son and stepson (who was Benjamin Franklin's grandson) in Neff's school and in 1824 recommended him for the position of director of Kentucky's state school system.[101] For Duane, there was "only one system of Education in existence fit for a country that is free, or for a people to whom intellectual knowledge is power and ignorance is weakness," and that system was the Pestalozzian system practiced by Neff.[102]

Neff's academy closed in 1813, and with its demise Philadelphia socialists turned again to economic affairs. In 1817, as the postwar depression began to deepen, a short pamphlet appeared in Philadelphia that explained the growing impoverishment of America's laboring classes in a new and trenchant way. Written by Cornelius Blatchly, a New Jersey physician and Christian socialist who would become, in 1829, the state assembly candidate of the New York Workingmen's party, *Some Causes of Popular Poverty* combined a radical theory of exploitation with a primitive, anti-denominational

Christian ethic.[103] Echoing Duane's radical reading of the small-producer tradition, Blatchly argued that "if wealth should produce opulence without the art, labour, or ingenuity of its proprietor, the opulent owner must necessarily obtain his increase from those who exercise art, labour, and ingenuity." Moreover, he declared, "if [the producer's] industry and labours are the *sole* causes of the opulence of nations, [then] they are the sole persons who ought to increase in opulence."[104] Here, then, were Duane's arguments recast in an open and more radical language. While Duane had broached the concept of exploitation in 1807, Blatchly turned it into the fundamental concept that would drive the formation of the early American working class: the idea that the producer alone had a right to the whole product of his labor.[105]

Blatchly advanced socialist theory in other ways as well. While Duane viewed taxation as the mechanism of surplus extraction, Blatchly saw the engine of exploitation as rent, interest, and the inheritance of property. "Rents of houses, and land, and interest of money," he declared, were "probably the effects of an ancient usurpation, tyranny, and conquest" stretching from the barbarians who sacked ancient Rome to the kings who plundered "the American aborigines of their lands."[106] Against property rights founded on these forcible expropriations, Blatchly proposed a concept of property more in tune with Old Testament principles. "Man," he wrote, "had *dominion* given to him, not in his *individual*, but in his *aggregate* capacity. If individuals usurp what is the divine right only of the aggregate, they deprive *man* (a term including all men & women) of his rights and privileges granted him in the beginning by God."[107] In light of this principle, Blatchly argued that "all men should . . . esteem themselves as deriving their titles from him for *general* use and benefit, and not for individual aggrandizement and oppression of the multitude."[108] Here then was the core of the small-producer tradition transformed into a socialist vision of Christian commonwealth. Against what he described as "this selfish, unfeeling, and misguided world," Blatchly offered a community of producers whose interests were mutual and whose labor was rewarded with competency and self-respect.[109]

In addition to men who drew their income from rents and interest, Blatchly assailed those who owed their wealth to the accident of inheritance. Inheritance, he wrote, was "more partial to the opulent than to needy or worthy people [and] more confined to men than women."[110] More than this, inheritance was a violation of the social nature of property. "Property is social," Blatchly argued, because it is "the result and fruits of social protection, policy and assistance." "If we owe so much to social union, and if our individual all is from it," he asked, "is not our individual all in a measure due to it? Does it not belong to it?" His answer was an unequivocal yes, and Blatchly went on to argue that individual property ought to revert to the social collectivity "as soon as death severs any individual of us from social rights and privileges."[111] If such an equitable policy were followed, he declared, then property could be distributed more equally among the productive members of society. Thus, unlike the grasping world of early nineteenth-century America, in Blatchly's Christian commonwealth every family would "be satisfied as having his average social property," and those who "have more than [their] due would be [seen as] an injury to others."[112]

In joining the elements of primitive Christianity with secular notions of expropriation and exploitation, Blatchly brought together two of the most powerful moral forces in early nineteenth-century America. Congregations of laboring-class Methodists, Baptists, Presbyterians, and Universalists had been exploring the "true" meaning of Christianity for more than a generation, and while their conclusions differed, they all agreed that God had not been well served by the selfish drift of American society.[113] Blatchly's Christian commonwealth provided a bridge between this Christian world of the streetcorner congregation and the radical world of the journeyman's society; and in so doing, he laid the foundation for a Christian-radical alliance against the profound immorality of the emerging social order. Given time to develop, this Christian-radical connection would play a critical role in the rise of artisan socialism and the workingmen's movement in the turbulent years ahead.

The youngest voice in Philadelphia's socialist community was also one of its most important. Stephen Simpson was a full generation younger than Duane when he joined the radical's inner circle. He had grown up in the shadow of the Bank of the United States where his father was chief cashier, and as a young man had himself become a note clerk in the Second Bank of the United States.[114] Then, in 1818, he abruptly resigned and exposed the incompetence and malfeasance of the bank's directors in the pages of the *Aurora*.[115] With these credentials, Simpson quickly became the Old School's preeminent political economist.

Simpson's writings were lessons for the workingman, and he aimed his messages carefully, weaving together the strands of the small producer tradition with the newer threads of radical political economy. Following Philadelphia's early socialists, Simpson began with the premise that "*Labor is the source of wealth*, and *industry* the arbiter of its distribution."[116] Yet as he looked around him, he found this principle everywhere violated. "They [who] do all the work, elect all the public functionaries, fight all our battles, gain all our victories, cause all our enjoyments to flow upon us," he noted, "still remain destitute of the frugal store of competence which ought to be the reward of industry."[117] Instead it was the nation's capitalists who "live and grow rich by the labour of others."[118] Anticipating Marx, Simpson explained that "Capital is naturally a tyrant; always standing on the alert to grind down the mere operative who lives from hand to mouth, and who must sell [his labor] because he must eat." There was "no other means than this fraud, monopoly, and unjust distribution of labour," he concluded, through which the capitalist could "grow rich and the industrious majority remain poor."[119]

Simpson's political economy was a critical solvent to the specious arguments of merchants, bankers, and speculators who sought to justify their massive accumulations of wealth by pointing to the service their capital performed for society. As his writings made clear, Philadelphia was now a capitalist society, and the contest was no longer one between producers and rich men who used their economic power to dominate society; now workingmen faced capitalists whose wealth came from their labor and whose power was exercised at their expense. Simpson's answer to the growth of American capitalism was a political organization of workingmen who would use the power of their suffrages to obtain "the true and just mode of distributing labor: by giving value for value."[120] This project would occupy Simpson and William Duane over the course of the coming decade and would eventually lead to the formation of the Workingmen's party in 1828.

The work of Simpson and Duane leaves little doubt that a radical tradition existed among Philadelphia workingmen that linked the eighteenth-century small-producer tradition with the nineteenth-century experience of industrial capitalism. Considerably refined since Sir William Keith's appeal to "the poor laborious part of mankind" in 1722, the small producer's creed was transformed by contact with Philadelphia's socialist community to become an increasingly militant set of ideas in the battle between craftsmen and capitalists. Moving from antipathy to a rising merchant aristocracy in the early eighteenth century to a justification for full political participation during the Revolution and then to the threshold of class consciousness in the early decades of the nineteenth century, the small-producer tradition was an intellectual and moral reservoir from which Philadelphia workingmen had drawn strength and sustenance for more than five generations.

But a moral tradition, the Old School leaders understood, was not the same thing as a social movement: missing was the organization necessary to translate the moral vision of a small-producer community into living reality. Since the end of the wartime rapprochement between Old and New School Democrats, popular politics in Philadelphia had been in disarray. In 1820, Nicholas Biddle, a dawning light of the New School, wrote that city politics was "in truth a perfect chaos of small factions."[121] Biddle's observation was born out by the November elections which confronted Philadelphia workingmen with a bewildering array of candidates representing Federalist, moderate Federalist, Old School, New School, and breakaway "Family" parties. Reflecting the increasing separation of financial, commercial, manufacturing, and working-class interests during the postwar years, the decomposition of local party politics was a further measure of the economic changes that were transforming Philadelphia from a commercial port into an industrial center.[122]

Lacking the personal connections and organizational skills of Michael Leib, who died in 1822, the Old School fell victim to the fragmentation of city politics. In the November 1820 election Old School candidates garnered only a third of the laboring-class votes in their traditional strongholds of Southwark and the Northern Liberties.[123] And two years later, running as a protest candidate in the Northern Liberties, William John Duane secured a mere 94 votes.[124] Much of the blame for the Old School's electoral decline can be traced not to a lack of commitment on the part of city workingmen, but to the political inexperience of Simpson and the younger Duane. In the fall of 1820 Simpson had successfully rallied a broad cross-section of Northern Liberties workers in a town meeting held to protest the neo-Federalist policies of President James Monroe, only to have the meeting broken up by a handful of Monroe's supporters.[125] Displaying an equal lack of political skill, W. J. Duane began his 1822 campaign so close to the October election date that he was unable to overcome the widespread sense of a certain Federalist victory.[126] Things looked brighter the following year when the Old School became the first political organization to champion Andrew Jackson's presidential candidacy, but by the 1824 election that advantage, too, disappeared as the better organized New School Democrats, now commonly referred to as the Family party, successfully claimed Jackson as their own.[127]

By 1824 it was apparent that the Old School was incapable of recapturing its past position as the pre-eminent party of Philadelphia workingmen. Hobbled by inexperienced leadership and weakened by the breakdown of their old multi-class coalition, the Old School party had in fact become more a forum for discussions between city radicals and workingmen than an effective political organization. William Duane might remain a thorn in the side of the Family Democrats, but his was a moral and not a political leadership. In 1823 John Lisle, a Family party leader, marveled that "at the ward meetings, and at town meetings, [Duane] can manage to bring forward such a number of the canail more than we can that they have often beaten us at these places."[128] But town meetings were one thing and organized coalition politics another. In city elections it was Lisle and the Family party, not Duane and the workingmen, who won.

CHAPTER 7

Making the Republic of Labor: The Workingmen's Movement and American Working-Class Consciousness, 1820–1830

Before 1820 socialism was an idea held mainly by middle-class radicals such as James Reynolds, William Maclure, and Cornelius Blatchly. William Duane and Stephen Simpson had popularized some of these ideas to be sure, but an artisan socialism had yet to manifest itself in Philadelphia. This would change in the 1820s as socialist ideas and communitarian designs flowered in the city of brotherly love.

Yet, at the same time, socialism was not the only body of moral thought competing for working-class allegiance. Since the turn of the nineteenth century, the proliferation of small, evangelical churches in the city's working-class districts had offered an alternative response to the changes brought about by capitalist industrialization. Emphasizing individual salvation at the expense of collective organization and substituting religious self-examination for socialist analysis, Philadelphia evangelists threatened to divide the working-class movement in its infancy.

In the end, the evangelical churches would attract no more than a minority of Philadelphia workingmen before the 1830s, but religion would play an important role in shaping the working-class movement nonetheless. By the middle of the 1820s, a young English-born shoemaker would fuse the elements of rational Christianity with the ideology of artisan socialism to create the nation's first workingmen's trade and political movement.

I

The socialist enthusiasm of the 1820s owed much to the communitarian vision of Robert Owen, an English manufacturer who had gained worldwide notoriety through his experimental industrial community at New Lanark, Scotland.[1] In New Lanark, Owen offered his factory operatives clean housing, good working conditions, and a measure of self-respect in an attempt to create an alternative model of industrialization devoid of the poverty, disease, and degradation that had become synonymous with England's industrial revolution. In 1817 Owen set out the ideas upon which New Lanark had been founded in a small book entitled *A New View of Society*.[2] Basing his arguments on the environmentalist psychology of David Hume, which held that human personality was in large part shaped by a person's surroundings, Owen claimed that given humane conditions and respectful treatment, industrial workers could become not only happy and efficient workers but respectable citizens as well. More of a communitarian than a socialist, Owen's vision had no room for the existence of classes or class conflict but projected instead a national community in which every citizen worked in his calling for the good of the whole. As Owen wrote in 1825, his object was "to change from the individual to the social system; from single families with separate interests to communities of many families with one interest."[3]

William Duane published Owen's *New View* in the *Aurora* during the early months of 1817, and from it, as well as from numerous English editions, small groups of people from the Philadelphia region began to take up Owen's communitarian idea. In the same year that Duane published Owen's tract, Thomas Branagan, a Methodist reformer and author of the early feminist tract *The Excellency of the Female Character*, gave favorable mention to Owen's ideas in his *Pleasures of Contemplation*, the work to which Cornelius Blatchly's *Causes of Popular Poverty* had been appended.[4] By 1823 two of the most respected leaders of the Academy of Natural Sciences, America's premier mineralogist Dr. Gerard Troost and the prominent

Quaker City druggist John Speakman, were actively planning an Owenite community outside Philadelphia.[5]

But while this publicity created an audience for Owenism, the major impetus to the spread of Owenite ideas in Philadelphia came from the arrival, in November 1824, of Robert Owen himself.[6] Owen was welcomed with uncommon enthusiasm by Philadelphia's reforming community, and his four days in the city were marked by a continuous round of brief speeches and public appearances. Owen's evenings were taken up with elaborate dinners given in his honor and with intimate conversations with the city's leading philanthropists. His guides throughout his visit were Richard Rush, the elder son of Dr. Benjamin Rush, and the publisher Mathew Carey, both of whom were prominent leaders of the New School wing of the Democratic party. His intense public schedule and the equally enthusiastic reception he received on his many subsequent visits to the city confirmed that "the remarkable Mr. Owen" was indeed the toast of reforming Philadelphia.

Yet while Philadelphia could credibly claim to be the seedbed of American Owenism, Owen's impact on the city's working classes is more difficult to measure. On the one hand, Owen organized and supplied his New Harmony, Indiana, community from Philadelphia, and in 1825 a group of city reformers planned to begin an Owenite community at Washington's wartime headquarters at Valley Forge. But these were distinctly middle-class communes with few mechanic members.[7] The fact that Owen consistently avoided the city's laboring-class suburbs and limited most of his public appearances to the Franklin Institute, a bastion of middle-class scientists and master manufacturers, suggests that Owen had little direct influence upon city working men. The great pains that the New School wing of the Democratic party took to escort Owen on each of his many visits to Philadelphia further suggests that his appeals were directed to manufacturers like himself rather than the humble mechanics of the city. If ordinary workingmen flocked to hear Robert Owen, there is no evidence to confirm it.

Owen's *ideas* did influence Philadelphia workingmen, however, although in an indirect way. During the 1820s a number of Owen's English followers began developing the political economy of his communitarian socialism. Foremost among these early English socialist economists were William Thompson, author of *An Inquiry into the Principles of the Distribution of Wealth* (1824) and *Labour Rewarded* (1827), and John Gray, whose *A Lecture on Human Happiness* (1825) made socialist economics available to the working classes.[8] Of the two, Thompson made the greater theoretical contribution, inventing the concept of surplus value upon which Marx would base his system a generation later.[9] But it was Gray who could communicate to the workingman, and while his theoretical contribution was meager, his influence upon Philadelphia's working-class movement was large. In 1826, as the workingmen's movement was beginning to form, Gray's *Lecture* went through three separate printings.[10] Two years later the *Mechanics' Free Press*, the nation's first labor newspaper and the official organ of the Workingmen's party, published the *Lecture* serially for its working-class readers. The influence of Gray's tract was, in fact, so profound that William Maclure credited it with awakening class consciousness among working-class Philadelphians.[11]

The uncommon power of Gray's *Lecture* rested in large part on its affinity to the main tenets of the small-producer tradition and the ideas put forward by Philadelphia radicals and socialists over the course of the preceding century. Beginning with the small-producer claim that *"society* is the natural condition of mankind," and that "the propensity *to exchange labour for labour"* was the foundation of human society, Gray went on to attack the eighteenth-century republican tradition for its notion that social evils stemmed from bad governments.[12] Why "do we so frequently attribute our miseries to the defects of governments," he asked, "since it is exclusively by barter that power is introduced into the world?"[13] All existing reforms, Gray argued, ignored the fundamental fact that the cause of human misery was economic rather than political in nature. "By an endless variety of charitable institutions, monuments equally to benevolence and ignorance, we attempt to subdue the evils of society; but the attempt is vain." Only solutions which *"annihilate the causes* whence the evils of mankind arise" and provide for "an equal distribution of the means of happiness to all" can guarantee success.[14]

From the beginning, Gray proposed not partial reform, but the total remaking of early capitalist society. The principles he defended spoke directly to the workingman. "Every necessary convenience, and comfort of life," he declared in an argument made familiar by Philadelphia's socialists, "is obtained by human labour [and] every member of the community who is not engaged in [labor] is an UN-PRODUCTIVE member of society" and hence exacts "a DIRECT TAX upon the productive classes."[15] In England, Gray calculated, the social producers received only one-fifth of the wealth they produced; the remainder went to non-producers such as manufacturers, who were "only useful as directors and superintendents of manufactories," and merchants, who were "mere *distributors of wealth, who are paid for their trouble by the labour of those who create it.*"[16]

Going on to examine the income of the non-productive classes more closely, Gray, like Blatchly, found that much of it came from rents and interest. For Gray, the practice of charging rent violated a fundamental principle of social justice. "The earth is the natural inheritance of all mankind," he maintained, and thus it belonged "to no man in particular, but to every man." "What does the landlord do?" Gray asked his readers, and supplied a self-evident answer: "He does nothing!"[17] Having thus expended no labor, Gray concluded, the landlord could claim no ethical right to the property he rented.

Likewise with interest. Following an argument popularized by Cornelius Blatchly a decade earlier, Gray characterized interest as yet "another mode of obtaining labour without giving an equivalent for it." But where Blatchly had rested his arguments against interest on a firm religious foundation, Gray turned instead to the labor theory of value. "All just contracts have for their foundation *equal quantities of labour*," he declared, but as the moneylender performs no productive labor, he therefore has no labor of his own to exchange. Thus, in the final analysis, the moneylender was able to lend money at interest only because his accumulated wealth gave him power over those in need of cash.[18] Like the traditional practice of artisans and workingmen, in a just society accumulated wealth would be used only for self-support in old age, not as a means to provide an income at another's expense. "If a man accumulates a fortune and chooses to retire upon it," Gray argued, "the moment he ceases to do something to support himself that fortune ought to decrease by every shilling he takes from it."[19] It was the only ethical thing to do.

In these and many other ways Gray's *Lecture* mirrored the arguments of the small-producer tradition as well as the moral imperatives of Philadelphia's early socialist writers. The radical extension of the labor theory of value, the antagonism between a large class of producers and a small, parasitic class of non-producers, and the idea of exploitation through surplus extraction were all part of Philadelphia's socialist tradition before the arrival of Gray's tract. To this Gray added an original theory of the pernicious effects of competition on working-class income and hinted at an under-consumption theory of capitalist crisis. Gray called capitalist competition "the fountain head of evil" that had "filled the earth with wretchedness, and baffled every attempt to render mankind virtuous and happy."[20] At its most apparent level, competition fixed "the quantity of wealth obtained by the productive classes" by driving wages down to a level "which is *just sufficient* to support bodily strength and to continue [the] race."[21] But, at a deeper level, competition also put a brake on social production itself. "In the present state of society," he declared, "production is limited by *demand*."[22] This being the case, low wages worked to lower social demand by limiting the income of working-class purchasers. And this, in turn, limited total social production and led to even lower wages and eventual unemployment. In this way, Gray told his working-class readers, capitalism led not only to a maldistribution of wealth but to an inefficient and crisis-ridden economy as well.[23]

In Gray's *Lecture* Philadelphia workingmen found a comprehensive argument for artisan socialism, and, more broadly, a socialism of the producing classes. In place of capitalist competition, with its chronically low wages, unemployment, and grinding poverty, producer socialism offered a cooperative community of farmers and craftsmen freed from the uncertainty of achieving a competency. As

Gray put it, "Let us abolish [competition] and we should then have as much wealth as we have the POWER OF CREATING!!!"[24] The small producer tradition had long held out the promise of just such a community, but it was John Gray and Philadelphia's early socialists who opened a new path toward its achievement. From the first mobilization of Philadelphia's laboring classes in the 1720s through the decline of Old School democracy in the second decade of the nineteenth century, Philadelphia workingmen had relied on an essentially political path toward a small producers' republic. But while democratic politics would remain a vital part of the city's working-class movement, after 1825 the economics of exploitation would become one of its driving forces.

II

While Gray's *Lecture* and the writings of the Philadelphia socialists prepared the way for artisan socialism, another crusade sought to engage the moral sympathies of city craftsmen. Religion had always been a part of artisan life, although it often took the form of a diffuse, non-denominational religiosity that historian John Bossy has labeled "traditional Christianity."[25] But as the Second Great Awakening spread from the rural hinterlands of America to the nation's seaboard cities at the turn of the nineteenth century, evangelical ministers began to recognize the vast, untapped potential of urban workingmen. Seeking to accumulate souls on a truly entrepreneurial scale, in the second decade of the nineteenth century these urban evangelists began an earnest campaign to tap the deep reservoir of religious sentiment that existed within Philadelphia's working-class community.

Philadelphia Methodists had, in fact, been preparing the way since the 1760s, when Thomas Webb, a British soldier turned itinerant, Edward Evans, a ladies' shoemaker, and James Emerson, a vendor of orange-lemon shrubs, first called their fellow craftsmen to worship in a sail loft along the Delaware waterfront.[26] Using the revival techniques perfected by George Whitefield during Philadelphia's first Great Awakening, Quaker City Methodists had, by the end of the eighteenth century, created congregations with a distinctively laboring-class hue. As early as 1794, for example, more than half of Philadelphia's white male Methodists were workingmen, while by 1801 craftsmen made up more than two-thirds of the city's Methodist congregations. Nor were the Methodists alone. In 1804 evangelical Presbyterians entered the competition for working-class souls by constructing a church in the Northern Liberties. Built to address the religious needs of the district's maritime craftsmen, by 1813 almost half of the suburban congregation was identifiably working-class.[27] While these evangelical congregations would remain small before the 1830s, it was nonetheless clear that in Philadelphia, workingmen and their families formed the bedrock of the evangelical movement.[28]

The success of these evangelical congregations among the city's working people was a tribute to the hard work and ecumenical doctrines of Philadelphia's evangelical ministry. At a time when middle-class congregations were creating new standards of moral rectitude and social propriety by separating themselves from the rough-and-tumble culture of city workingmen, urban evangelists actively sought out working-class converts. The evangelists' heartfelt concern for the spiritual well-being of ordinary workingmen and their families established a powerful link between these ministers and their working-class congregations, links that threatened to pull many converts away from the traditional working-class community altogether.

But beyond these personal linkages and the simple cure of souls, city evangelists offered a message of collective redemption that had not been heard in Philadelphia since George Whitefield's mid-eighteenth-century revivals. The reigning Protestant doctrine of predestination—a belief in the divine election of a handful of "saints" and the eternal damnation of all others—had lost much of its moral force by the turn of the nineteenth century. In fact, to many Americans the idea of eternal damnation itself ran counter to the republican and democratic notions of the age. Sin abounded in the hearts of men, to be sure—here was a point on which all evangelists agreed—but grace was God's free gift to every sinner. Methodist and Presbyterian evangelists hammered on this point again and again: seek salvation with earnest diligence and abundant prayer, and it will surely come. The key to salvation, they told their congregations, was a recognition of human sinfulness coupled with the necessity of complete submissiveness to the will of God. If they accepted these simple principles, the ministers assured their followers, they would truly be born again.

Yet, while clerical empathy and democratic doctrine played an important role in winning working people to the evangelical cause, the real power of evangelicalism lay in its emotional connection with existing artisan religiosity. The tradition of artisan dissent which began in sixteenth-century England and found its most vocal expression in the trans-Atlantic revivalism of the mid-eighteenth century had taught artisans to demand an intensely personal and emotional religious experience. As one early nineteenth-century Methodist recalled, people expected "tornadoes" and if a preacher "did not break down everything before him," he would not long have a congregation.[29] Sometimes this was meant literally, as when the flamboyant New York itinerant "Father Turck" would "clap his hands, and lift up his chair and dash it down on the floor, and call for the power until he made everything move."[30] More often it meant a class meeting where everyone made a "considerable noise" in praying, singing, and shouting for redemption.[31]

Here, then, was the secret of evangelical success in early nineteenth-century Philadelphia. For some city workingmen, evangelicalism held forth the prospect of a new type of community, one that could encompass both the emotional anxiety of uncertain times and traditional artisan concerns about competence, independence, and community respect. Class meetings, love feasts, and spontaneous testimony—the social apparatus of evangelicalism—gave workingmen and their families the freedom to express their deep-felt apprehensions about the declining craft system at the same time that the congregation itself provided a positive experience of solidarity and mutual assistance not unlike that found in the traditional craft community. In the end, a number of workingmen felt, perhaps the sense of personal connection and collective mutuality that was absent in the new capitalist workplace might be recovered in the electric atmosphere of the evangelical congregation and prayer meeting.

A nineteenth-century mariner described the emotional energy of the evangelical appeal with great poignancy: "What I likes along o'preachin'," he told a traveling evangelist, is "when a man is a-preachin' at me I want him to take somert hot out of his heart and shove it into mine, —that's what I calls preachin'."[32] It was by creating a profoundly *personal* experience of salvation that both encompassed and redefined traditional working-class notions of community and religious life that urban evangelists drew artisans into the evangelical fold. By offering a collective setting for an otherwise solitary religious experience, evangelicalism created an alternative community that drew its power from the traditional artisan community at the same time that it formed its very antithesis. By redirecting popular fears and religious sentiments into rigid denominational channels, the evangelical movement created communities of faith potentially as powerful and cohesive as the artisan communities which were their most formidable competitors.

As Philadelphia's working-class leaders understood, in the act of forging its own community bonds, evangelicalism drove a wedge into the city's artisan community. The easy coexistence of small-producer and artisan religious traditions that had characterized the seventeenth and eighteenth centuries came apart during the 1820s, driven asunder by the force of ideological competition. In the end, the competitive denominationalism of the city's Methodist and Presbyterian evangelists threatened to divide the working class along religious lines and dilute the power of artisan socialism and the workingmen's movement in the process. Although the full force of evangelicalism would not be felt until the 1830s and 1840s, the divisive influence of Philadelphia's evangelical communities cast a shadow across the 1820s, threatening the very success of the workingmen's project.[33]

III

By the mid-1820s Philadelphia socialists had developed a powerful critique of early capitalist society that paralleled the tenets of the small producers' creed. All that was lacking was a solid link to the city's expanding and increasingly militant journeymen's societies. No more than a small handful in number at the turn of the century, journeymen trade societies grew rapidly during the 1820s.[34] Although the lack of surviving organizational records makes an exact count impossible, notices in the city newspapers and local directories reveal that journeymen hatters, coopers, ship carpenters, handloom weavers, tobacconists, ladies' cordwainers, bricklayers, millwrights and machine makers, blacksmiths, whitesmiths, saddlers, and harness makers joined the previously organized printers, tailors, cordwainers, house carpenters, cabinetmakers, curriers, stone cutters, and mariners in forming independent trade societies.[35] Even more dramatic than this growth in organization, however, was the unprecedented militancy of the societies themselves. Before the second Anglo-American war Philadelphia employers seldom faced their journeymen as strikers; in fact, the first fifteen years of the nineteenth century recorded only three turnouts by journeymen shoemakers, two by printers, and one by curriers.[36] The immediate postwar years saw no recorded strikes as the depression and Panic of 1819 brought large-scale unemployment and a weakening of the journeymen's bargaining position. But with the return of relative prosperity, journeymen ship carpenters, hatters, weavers, cabinetmakers, house carpenters, tailors, bricklayers, and cotton spinners carried out ten strikes against their employers in 1821 alone.[37]

Both the growing number and uncommon militancy of the trade societies during the 1820s suggest that a fundamental change was taking place in the minds of Philadelphia's journeymen. The normal reflex has been to explain this change by reference to the economic consequences of capitalist industrialization: the effects of widening markets, improved transportation, the introduction of new methods of production, and the use of convict labor.[38] But while these changes were important, the behavior of Philadelphia journeymen was more than a narrow reaction to economic change. These were, after all, the same men who had been the mainstay of Old School Democracy and had met with Michael Leib and William Duane in neighborhood taverns for more than a generation.

It was to these men, journeymen in and out of trade societies, that

William Heighton directed his call for a working-class movement in Philadelphia.[39] Coming to the city as a youth just before the second Anglo-American war, Heighton was apprenticed to the shoemaking trade that dominated life in his native Northamptonshire.[40] Little is known of his early adulthood, but by 1826 Heighton had absorbed the socialist writings of John Gray and was actively canvassing the city's laboring-class suburbs with his message of Christian brotherhood, working-class unity, and artisan socialism. He spoke to an audience already well prepared.

The problem faced by workingmen, Heighton declared, was their lack of unity in the face of a powerful class of non-producers. What was needed, he argued, was a union of trade societies joined to an artisan-based political party. William Duane had long urged Philadelphia craftsmen to combine trade organization with party politics, but it took the energies of a fellow artisan like Heighton to unite city craftsmen into a working-class movement. Heighton spent the fall and winter of 1826 discussing his ideas among friends and fellow journeymen throughout the city, and by early 1827 he was ready to test his plan in the public arena through a series of public addresses. Drawing heavily on Gray's *Lecture on Human Happiness*, Heighton brought the disparate strands of the small-producer tradition, Philadelphia socialism, rational Christianity, and Old School democracy together into an economic and political prescription for small-producer socialism in the Delaware city.[41]

In his *Address to the Members of Trade Societies*, Heighton set the terms of his program. Following the radical reading of the labor theory of value pioneered by Duane and the Philadelphia socialists, he pointed out that working people were "the sole authors of all the luxuries . . . and of all the property or wealth that is in existence." But, he continued, even a brief survey of the condition of the city's producers revealed that they were "put off with a scanty portion of the coursest and meanest of their productions [and were] deprived of almost everything which is calculated to . . . render life a blessing."[42]

The cause of this imbalance in Philadelphia's moral economy was well known: it was competition and the exploitation of the producing by the non-producing classes. Following Gray, Heighton placed

Philadelphia's non-producers into six classes—legislative, judicial, theological, commercial, independent, and military—each of which extracted a portion of the producers' labor for their support.[43] But while each of these classes lived on the labor of workingmen, even more pernicious was the "system of competition" which set man against man. Here Heighton quoted the heart of Gray's *Lecture*, a chapter entitled "Competition the Limit to Production," in its entirety, closing with his own prediction that "half of the miseries which [competition] is capable of inflicting upon society have not yet been told."[44] Thus for Heighton, as for Gray, Simpson, and the Philadelphia socialists, parasitic non-producers and capitalist competition were the source of the degraded state of the American working class. The prospect before them "as a class" under the present system, he warned, was a "gloomy one of endless toil and helpless poverty." Unless, that was, "we now begin and prepare *ourselves* to obtain and enjoy an equal portion of the immense benefits that wealth can confer."[45]

The process of restoring a measure of justice and equality to social relations in Philadelphia required two things, according to Heighton. First, workingmen needed to recognize the real social value of their labor and the ways in which non-producers extracted that labor for their own use. Only with this knowledge clearly before them, he argued, could workingmen unite across craft lines and utilize their combined strength to achieve the most important objective, political power and the restoration of the producer to his rightful place in society. "Surely we, the working class," he declared, "who constitute a vast majority of the nation [and] who create all the wealth that is created, have a right to expect [that] an improvement in our *individual condition* will be the natural result of [all] legislative proceedings."[46] Yet, wherever one looked, it was the merchants, manufacturers, and western land speculators rather than the mechanics or farmers who ruled. By 1824, even President James Monroe and the Democratic Congress were instituting national tariff and banking policies that were Federalist in all but name. Given the political realities of the 1820s, Heighton told his fellow workingmen, they could expect little from their elected representatives.

The problem was the caucus system. Since the end of the Revolution, voters in Philadelphia had been presented with a list of candidates selected not by the people themselves, but by an inner circle of party leaders. As William Duane had been arguing for more than a decade, the caucus system effectively isolated legislators from the people at large and in the process created a class of politicians more attuned to their own political fortunes than the interests of the people they were elected to represent. To counter the pernicious effects of the caucus system, Heighton resurrected an idea put forward by Philadelphia artisans in the 1780s and called upon craftsmen to nominate and elect their own candidates. "I think it pretty clear," he observed, "that we shall never reap any benefit from [politicians] until we have men of our own *nomination*" in the legislature, "men *whose interests are in unison with ours.*"[47] Only then, he concluded, would Philadelphia producers "have REAL REPRESENTATIVES, and a public opinion . . . through which to direct and control them."[48]

With Heighton's speeches we are firmly on the ground of class consciousness: working people produce society's wealth and receive few of its benefits; merchants and manufacturers extract a portion of this wealth "under the name of PROFIT" which becomes, for the working class, "an iron chain of bondage."[49] One has only to add organization and one has, full-blown, a working class. It was altogether fitting that a working journeyman should be the one to bring English radical economics, Old School politics, and Philadelphia's socialist tradition together for his fellow workingmen.

For the next two years Heighton stumped through the city, calling on local trade societies, visiting workers in their homes, sharing drinks in neighborhood taverns, and making his speeches to packed audiences in the Northern Liberties and Southwark.[50] His message to workers was threefold: study political economy in order to understand the source of your exploitation, organize, and act to create a just and equitable society.[51] To help workingmen obtain "needful intelligence," he proposed "the establishment, in every city and large town, of a Library . . . for mutual instruction [and] a POOR MAN'S PRESS . . . appropriated to the interests and enlightenment of the working class and supported by themselves."[52] To give journeymen economic power and a greater reward for their labor, Heighton also proposed the creation of a powerful central association of Philadelphia's journeymen's societies that would collect and administer a joint strike fund, direct turnouts, and recruit new members. Such an association, he thought, would give to Philadelphia's wage-earning journeymen a collective power they could not hope to have in their independent trade societies. With the reins of economic power restored to the hands of the city's producers, Heighton thought, the association would be better able to support a political branch which could overcome the system of special privilege that ruled Pennsylvania politics and restore true democracy to Philadelphia.[53] "If the members of the different Trade Societies in this city are willing to follow this program," he declared, "I see nothing which can materially impede their progress or prevent their success."[54]

Throughout these long months of organizing, Heighton relied heavily on the assistance of the city's two Universalist churches. Already a meeting-ground for Philadelphia radicals in the early days of the Democratic Society, the Universalists made it a policy to open their halls to grassroots organizations and ad hoc meetings of city workingmen.[55] Some of Heighton's most important speeches were, in fact, delivered in the Southwark church near his home.

But beyond creating a venue for Heighton's oratory, the Universalist churches played a crucial moral role as well. In an address delivered at the Southwark church in 1827, Heighton proclaimed the rational anti-Calvinism of the Universalists to be the moral handmaiden of the city's emerging working-class movement.[56] With its rejection of eternal damnation as a cruel and unenlightened doctrine and its uniquely humanistic vision of an understanding and benevolent God, Universalism attracted a large laboring-class following from the start.[57] As early as 1793, almost two-thirds of the communicants of Philadelphia's First Universalist Church were men with laboring-class occupations, and by the 1820s more than three-quarters of the membership came from the ranks of the city's working class.[58]

Much of this working-class appeal came from the close correspondence between Universalist doctrines and the mutuality and rational outlook of traditional artisan culture. Unlike city evangelicals, who appealed to the emotional side of working-class life, Universalism appealed to its rationality. An early Universalist hymn put it this way:

> My body must I under keep,
> Subjected to my mind;
> Lest like a lost and wand'ring sheep,
> Destruction's road I find.[59]

For the Universalists, God's moral universe was ordered and rational thought was the path to redemption. But, at the same time, the Universalists also recognized that men were far from perfectly rational creatures. The Universalist God expected transgressions and understood them sympathetically as part of human nature. "What is sin?" the Universalists asked, "is it not want of holiness?" "Will [God] say that man's to blame [for] obeying nature's laws?"[60]

Finally, against the evangelical emphasis on self-denial and the strict observation of denominational rules, Universalists emphasized free inquiry, self-improvement, and a community of rational love.

> May all the sons of Adam's race,
> Their ev'ry faculty improve,
> Till discord thro the world shall cease,
> And ev'ry creature meet in love.[61]

In Heighton's hands, this appeal to Christian mutuality was made to extend beyond the boundaries of church fellowship to embrace the workingmen's movement itself. The moral purpose of religion, Heighton declared, was the same as that of the workingmen: "to teach the absolute necessity of undeviating justice between man and man."[62] Accordingly, Heighton called for an alliance between the city's authentically Christian clergy and the fledgling workingmen's movement. Only by uniting with the workingmen, Heighton argued, could Christian clergymen hope to turn back "the system of individual interest and competition" that was relentlessly eroding the moral fabric of Quaker City life.[63]

Yet, while Heighton found Philadelphia's Universalist clergy sympathetic to the workingmen's cause, the response of the city's established clergy was a different matter. True Christian ministers ought to be living "imitators of those primitive Christians who had 'all things in common,'" Heighton declared, but in Philadelphia he found that most established clerics remained blind and indifferent to the "legalized extortion" practiced on the city's working classes.[64] "Why do they not point out the enormous injustice of one class of men possessing legal authority to take advantage of the necessities of another?" he asked.[65] And why hadn't the denominational clergy directed "the power of their reasoning, and the thunders of their eloquence against the unjust and vice-creating system of conflicting interest—a system so directly opposed to that adopted by . . . the Prince of Peace?"[66]

If the established clergy were culpable for their failure to speak out against this pernicious evil, the evangelicals were, for Heighton, the most culpable of all. Taking aim at the city's evangelists who were attempting to build a pious alternative to the workingmen's movement, he pointed to the futility of their quest for working-class redemption in a world structured by industrial capitalism. "Not all the fervent intercessions of prayer, not all the influence of pathetic exhortation, nor all the declarations of divine denunciation, can ever arrest the progress of sin while the system of individual interest and competition is supported," Heighton warned.[67] For him, the only hope for redemption lay not in working-class piety and accommodation with the present system, but in a united effort by clergy and workingmen alike to overturn that system in the name of social justice.

Religion, then, was closely bound up with Heighton's vision of a republic of independent producers. For him, the competition and individual interest of the wage-labor market was more than an economic and moral injustice, it was also a sin. But if the interest of the clergy and the interest of workingmen were the same—the creation of a moral society founded in social justice—the clergy would have to do more than "teach evangelical truths."[68] "If the clergy would arrest the fatal march of vice," Heighton advised, "let them direct their attacks on its fountain head."[69] "The grand nursery of sin must be destroyed," he added, "before they can cherish any reasonable hopes of a general and permanent reformation in our country."[70] For Heighton, the republic of labor would be more than a society of loosely associated producers, it would also be God's true community.

With the able assistance of these Christian precepts and the venue of Universalist meeting halls, by the fall of 1827 Heighton had the support of sufficient numbers of Philadelphia journeymen to put his plan into effect. Together with craftsmen from a cross-section of the city's organized trades, Heighton formed the Mechanics' Union of Trade Associations, to afford, its constitution declared, "each other mutual protection from oppression." They had long been suffering from the evils of "poverty and incessant toil" as the result of "an unequal and very excessive accumulation of wealth and power [in] the hands of a few," the mechanics continued, and were now joining together to redress this shameful state of affairs.[71] "Do not all the streams of wealth which flow in every direction and are emptied into and absorbed by the coffers of the unproductive, exclusively take their rise in the bones, marrow, and muscles of the industrious classes?" the journeymen asked the public. "Is it just," they concluded, that in return for their "unceasing exertion and servility" they received only "a bare subsistence (which likewise [was] the product of their own industry)" and nothing else?[72]

The Mechanics' Union thought not. In a remarkable restatement of the central tenets of the small-producer creed, their constitution declared that "the real object" of the association was to stop the "depreciation of the intrinsic value of human labor," to raise "the productive classes to that condition of true independence and equality which their practical skill and ingenuity, their immense utility to the nation" demanded, and, finally, to promote "the happiness, prosperity, and welfare of the whole community."[73] Here, in 1828, was compelling proof of the power and continuing relevance of the small-producer tradition, a set of moral precepts that had first appeared, three hundred years before, among the artisans of early modern England.

In May 1828 Heighton began publishing the *Mechanics' Free*

Press, America's first working-class newspaper. Created to guide the Mechanics' Union and Philadelphia journeymen in general, the paper reported the activities of the Mechanics' Union, advertised meetings, and informed its readers of the activities of workingmen throughout the nation. The main purpose of the paper, however, was to carry forward the education of the city's working class. In the three and a half years of its existence, Philadelphia journeymen could read in the *Free Press* a complete reprint of Gray's *A Lecture on Human Happiness*, articles on Pestalozzian educational reform, and Heighton's own editorials on artisan socialism. There, too, journeymen were first exposed to the idea of producer and consumer cooperatives, a project they would turn to increasingly in the decades ahead.[74] Thus, like Heighton's lectures, the *Free Press* brought together the separate strands of the small-producer tradition, Philadelphia socialism, and the politics of Old School Democracy.[75]

With the formation of the Mechanics' Union and the educational work of the *Mechanics' Free Press* well under way, Heighton next turned his attention to the politics of artisan socialism. "A new distinction of parties is about to originate," he announced in the summer of 1828, "involving, on the one side, the *Industrious Classes*, and on the other, the *idle and unproductive*."[76] The "new standard of politics" to which he referred was the world's first Workingmen's party. Founded at a meeting of the Mechanics' Union in early 1828, the Workingmen's party was designed to overcome the monopoly that Family party Democrats exercised over local politics and to offer a political alternative to the city's working class.[77] "The truth is," Heighton admitted to a meeting of Southwark journeymen, "the Working People of this country have never yet been *faithfully* represented in their legislative councils by those to whom they have given *their* suffrages." Instead, "we have permitted the wealthy and the proud, whose INTEREST it has been, and is, to render us poor and degraded, to NOMINATE, as candidates for public offices, individuals of their own stamp."[78] The result had been the system of "charters, statutes, and enactments, passed from session to session [of the legislature], for the exclusive advancement and benefit of Banking, Insurance and Mercantile, *master* Manufacturing,

Landed, and other monied, monopolizing, and speculating institutions and interests." Rather than allowing non-producers to continue their use of "the fostering wing of legislative protection [to accumulate] their annual millions from the toils and labours of the Operative classes," the Workingmen's party promised to represent the true interests of city producers.[79]

In its platform, the Workingmen's party called for broad reforms including free public education, the creation of a mechanic's lien law, the abolition of imprisonment for debt, and an end to the caucus system of nominating candidates. With the election of small-producer representatives and the enactment of these reforms, party leaders claimed, Philadelphia craftsmen could at last begin the long task of creating a small-producer republic. With the collective power of Philadelphia producers united in the workingmen's movement, what was lost by the post-Revolutionary generation of radicals and mechanics would be regained by their grandsons.

Taken as a whole, the platform of the Workingmen's party reflected the powerful influence of William Duane and Old School Democracy on the early working-class movement. Each of the reforms called for by the party in 1828 had, in fact, been part of the Old School program since Jefferson's first term. The caucus system had been an Old School target since the first decade of the nineteenth century, when Duane began a series of editorials urging a system of popular nomination and direct small-producer representation to replace the old elite practice. More recently, Michael Leib had introduced a bill for the abolition of imprisonment for debt in the 1817 state assembly, while Old School assemblyman James Thackera had sponsored a bill for free public education in 1819.[80] While none of these efforts proved successful at the time, they were remembered, and formed the core of working-class politics well into the 1840s.

The connection between Duane's small-producer ideals and the Workingmen's party went well beyond a common legislative program, however; it involved the direct participation of the Old School leaders as well. In 1828, and again in 1829, William Duane and his son served as prominent members of the party's banking committee, helping to shape both its anti-banking views and its general economic policies.[81] Then, in 1830, Stephen Simpson, an early party supporter as well as the Old School's economic theoretician, became the Workingmen's candidate for Philadelphia's congressional seat.[82] Thus, like the workingmen's movement itself, Philadelphia's Workingmen's party was less a singular innovation than the culmination of long years of learning and struggle on the part of radicals and workingmen alike.

The brief history of the Workingmen's party is a well-known feature of early Jacksonian politics, and only its main lines need be rehearsed here. After its first electoral contest with the Democratic party in 1828, the Workingmen faced an unpromising future. Of the thirty-nine candidates it fielded for state and national offices, only those who ran as joint candidates on other tickets were elected. The fate of those who ran alone on the Workingmen's ticket was simply an embarrassment: the most successful candidate received a mere 539 votes, less than one-tenth that of his Democratic rival.[83] Heighton dismissed this poor showing as the inevitable birth pangs of a new political party, and, in part, he was right.[84] But a more important cause was undoubtedly the popularity of Andrew Jackson, who carried Philadelphia by a large plurality.[85] The truth was that in 1828 many Philadelphia workingmen were still hopeful that the election of Jackson and his fellow Democrats would herald a new era of economic and social justice for American producers. But by the summer of 1829, as Jackson made the last of his federal appointments, even the most meager hopes of city craftsmen were dashed. Ignoring his closest advisors, Jackson appointed Family party men to the most important Pennsylvania posts and ignored his earliest and most ardent supporters.[86] Among those passed over was William John Duane, whom Jackson had originally intended to appoint as the federal district attorney for eastern Pennsylvania in recognition of his father's early support of his candidacy.[87] Disgusted by Jackson's appointments and his evident betrayal of the workingman, workingclass Philadelphians flocked to the Workingmen's party, which had organized energetic ward committees in the wake of the 1828 defeat.

The results of the 1829 election were dramatically different from those of the previous year. In 1829, the voters elected sixteen

Workingmen's party candidates and also installed all but one of those candidates who ran on multi-party tickets. In all, the Party polled close to 2400 votes in the city and county, nearly three times the number received the year before. Heighton was jubilant: "The balance of power has at length got into the hands of the working people, where it properly belongs," he declared, "and it will be used, in the future, for the general weal."[88] Although the "balance of power" held by Philadelphia's working class would shortly be used in ways he had not foreseen, Heighton's assessment of the political situation in the city was notably sound. In 1829, the Workingmen's Party did hold the balance of power in Philadelphia politics, a fact attested to by the serious efforts of the Democratic party to win working-class votes and to include working-class candidates on their subsequent city tickets.

In the end, the Workingmen's party met the fate of all third parties in American politics. After 1829, city workingmen increasingly turned to the Democratic party in order to utilize its power and plentiful resources for the benefit of the producing classes. Viewing the Workingmen's overwhelming success in the 1829 election with alarm, the mainstream Democratic party moved quickly to incorporate Workingmen's reforms into their platform, and in the course of the 1830s established free public schools, abolished imprisonment for debt, and staunchly supported the working-class ten-hour movement. Having learned a painful lesson in 1829, the Democrats would never again ignore the interests and desires of city craftsmen.

As for the Workingmen's party, the balance of power that Heighton had celebrated in 1829 lasted less than a year. The party did well enough in the 1830 election, increasing its number of voters by one-seventh and electing eight county commissioners in the Northern Liberties.[89] The party even played a prominent role in defeating mainstream Democratic Congressman Daniel H. Miller, president of the Bank of Penn Township and a symbol of the endemic corruption of Philadelphia's Jacksonian party machine.[90] But, despite these successes, the Workingmen's party proved no match for the revitalized Democrats, who garnered a commanding two-thirds of the 1830 vote and gained a hold on Philadelphia politics that would last until the early 1840s. In the 1831 election the party mounted a desultory campaign and, failing to elect a single candidate, passed from the political stage altogether.[91]

With the fall of the Workingmen's party William Duane passed into well-deserved retirement while his son, William John, achieved brief notoriety when, as Jackson's Secretary of the Treasury, he refused to comply with a presidential order to remove federal deposits from the Bank of the United States. Stephen Simpson attracted little more than protest votes in his 1830 congressional campaign and thereafter played a less exalted role as a minor functionary of the Democratic party. As for William Heighton, the creator of the Workingmen's party followed his project into oblivion, leaving the city for Indiana sometime after 1832 and emerging only once, in 1865, to write in favor of radical reconstruction and the right of former slaves to confiscated land and the undivided fruits of their labor.[92] All that remained was for contemporaries to divine the meaning of the workingmen's movement. It is a project that historians have continued to this day.

EPILOGUE
After the Workingmen's Party:
Labor and the Politics of Class

It has always been easy to dismiss the workingmen's movement as the naïve brainchild of middle-class reformers or the misguided actions of craftsmen smitten with the apparent efficacy of politics. Yet it was neither of these. The formation of the Mechanics' Union and Workingmen's party marked the birth of a working class. The fact that neither the union nor the party survived to lead the working-class movement through the 1830s was itself unimportant; for the issues raised by the workingmen's movement continued to frame the agenda of union leaders, working-class reformers, and party politicians for generations to come.

If the participants in this early working-class movement have often been seen as quaint and unsophisticated "babes in knowledge," it has been the purpose of this book to present an alternative view.[1] The workingmen's movement was the culmination of a century of intellectual and political struggle through which Philadelphia artisans fought to create an independent place for themselves in an emerging industrial world. At the heart of that struggle was a recognition that class and politics were inseparably connected and that the values of the past could be preserved only by collective action in the present. The economism of the twentieth-century working class was not the norm of their early nineteenth-century predecessors.

There were disagreements, to be sure, and in the wake of the workingmen's movement many Philadelphia labor leaders adopted a stance of studied cynicism toward organized politics that often sounded like anticipations of the "pure and simple" trade unionism of the American Federation of Labor. But despite this loss of faith on the part of many labor leaders, rank-and-file craftsmen were much less willing to abandon one of their most powerful traditions. No matter what they thought of the workingmen's movement, working-class Philadelphians embraced the possibilities of politics, in part because a century of experience had taught them to think in predominantly political idioms, but also because government impinged on so much of their daily lives.

This rethinking of the politics of class dominated working-class activity in the years immediately following the fall of the Working-men. As early as 1834, delegates to the National Trades Union (NTU) convention found themselves confronting the issue of political involvement from the start. On the second day of the convention, Robert Townsend, a New York City house carpenter, called for the creation of a committee to report on "the social, civil, and political condition of the laboring classes of the country."[2] Almost immediately, Howard Schenck, representing the Newark shoemakers, objected to the use of the word "political," not because he was against political action, but because "it was misunderstood by many" and might "array . . . against them the force of one or other of the political parties."[3] In the ensuing debate, William English and John Ferral, two of Philadelphia's most influential labor leaders, framed the issues of debate and testified to the continuing centrality of politics in the early working-class movement.

Representing the journeymen cordwainers, William English argued against any involvement in politics, citing the fate of the Mechanics' Union five years before. From the moment the union involved itself with politics, English claimed, "the Union retrograded" and collapse came by the end of the year. "The same cause would produce the same effect" for the NTU, he warned, and urged that the convention "avoid everything that might have a tendency to prevent a general Union of the Trades."[4] Politics was the domain of self-interested party men, English argued, and had no place in the working-class movement.

John Ferral, representing the Manayunk operatives, took a more pragmatic view. He, too, did not want the working-class movement divided over the use of a simple word; but "if the delegates thought their constituents were only babes in knowledge," he suggested sarcastically, "why not feed them with pap, and as they advanced give them more solid food?" For the sake of unity, he proposed that the convention substitute the word "intellectual" for "political," although, he pointed out, "he was fully satisfied that the working classes would never effectually remedy the evils under which they were suffering until they carried their grievances to the polls."[5]

Ferral's amendment carried the convention, but it is his tone which was more important for the future of Philadelphia's working class. Politics might be divisive, or even misunderstood, but it was essential to the success of the working-class project. As Boston delegate Charles Douglass put it, workers "had become degraded by bad legislation; they had got into difficulty by it, and how were they to get out but by legislating themselves out?"[6] After all, Robert Townsend pointed out, the working classes "hold the balance of power, and it was only necessary for them to say what they want, and each party would be anxious to adopt their measures."[7] The same point had been made by Heighton in 1829, and even the skeptical William English had to admit to the convention that the objectives of the Mechanics' Union "could not [have been] obtained without its taking part in politics."[8]

As the debate at the NTU convention makes clear, Philadelphia's labor leaders might disagree about the role of politics in the working-class movement, but even the staunchest supporters of economism had to admit the inevitability of political confrontation in the new manufacturing world. After 1831, politics continued to be the battleground on which the war between the "producing" and "accumulating" classes was fought. Unions remained vital to the movement, to be sure, but free public education, the end of debtor's prisons, and the ten-hour workday—*the* issues of the 1830s—could be realized only through political action. Like the Workingmen's party before them, the working-class movement of the 1830s had something larger than their wages and working hours in mind. If a revival of the craft system was doomed after 1831, the quest for independence would remain. And in the democratic world of Jacksonian America, it was impossible to think of independence without thinking first of politics. This was perhaps the greatest legacy of Philadelphia's first working class.

Notes

PROLOGUE

1. William Heighton, *An Address to the Members of Trade Societies, and to the Working Classes Generally* (Philadelphia, 1827), 12.

2. Among the most recent works to adopt this view are Sean Wilentz, *Chants Democratic: New York City and the Rise of the American Working Class, 1788–1850* (New York, 1984); Bruce Laurie, *Working People of Philadelphia, 1800–1850* (Philadelphia, 1980); and idem, *Artisans into Workers: Labor in Nineteenth-Century America* (New York, 1989).

CHAPTER 1
ANGLO-AMERICAN TRADITIONS

1. *Pennsylvania Gazette* (Philadelphia), May 12, 1768.

2. Ibid.

3. Ibid.

4. John Treat Irving, *An Oration Delivered on the 4th of July* (New York, 1809).

5. Walter Brewster, "The Mechanick on Taxation," *Norwich Packet* (Norwich, Conn.), April 4, May 10, 1792.

6. Ibid.

7. For similar attitudes among English artisans, see John Rule, "The Property of Skill in the Period of Manufacture," in Patrick Joyce, ed., *The Historical Meanings of Work* (New York, 1987), 99–118.

8. Workshop relations remain largely unexplored for the period before the 1840s. W. J. Rorabaugh, *The Craft Apprentice: From Franklin to the Machine Age in America* (New York, 1986), and Robert Bruce Mullin, ed., *Moneygripe's Apprentice: The Personal Narrative of Samuel Seabury III* (New Haven, Conn., 1989), 54–81, provide two impressionistic accounts.

9. Competency was also the goal of early American farmers. For an illuminating discussion of competency in rural New England, see Daniel Vic-

kers, "Competency and Competition: Economic Culture in Early America," *William and Mary Quarterly* 3d ser., 47 (1990), 3–29.

10. Thomas Dilworth, *A New Guide to the English Tongue* (Philadelphia, 1791), 120.

11. The transition from feudalism to capitalism has been the subject of historical debate for much of this century. The best recent account of this process is Robert Brenner, "Agrarian Class Structure and Economic Development in Pre-Industrial Europe," *Past and Present* 70 (1976), 30–75, and his reply to critics, "Agrarian Class Structure and Economic Development in Pre-Industrial Europe: The Agrarian Roots of European Capitalism," *Past and Present* 97 (1982), 16–113. Also important are R. H. Tawney, *The Agrarian Problem in the Sixteenth Century* (New York, 1967, 1912); Joan Thirsk, ed., *The Agrarian History of England and Wales*, IV (London, 1967); Maurice Dobb, *Studies in the Development of Capitalism* (New York, 1947); and R. H. Hilton, *The Decline of Serfdom in Medieval England* (London, 1969).

12. On the thoughts of craftsmen in the clothing districts, see Christopher Hill, *The World Turned Upside Down: Radical Ideas During the English Revolution* (New York, 1972), 22, 34, 78; A. G. Dickens, *Lollards and Protestants in the Diocese of York, 1509–1558* (Oxford, 1959), 48–50, and ch. 2; and J. F. Davis, "Lollard Survival and the Textile Industry in the South-east of England," *Studies in Church History* 3 (1966), 191–201. On London, see Brian Manning, *The English People and the English Revolution, 1640–1649* (London, 1976), chs. 9 and 10, and Margaret James, *Social Problems and Policy During the Puritan Revolution, 1640–1660* (London, 1930), ch. 5.

13. See Hill, *World Turned Upside Down*, 19; A. G. Dickens, *The English Reformation* (New York, 1964), 26–33; idem, *Lollards and Protestants*, 8; Davis, "Lollard Survival"; and Keith Thomas, *Religion and the Decline of Magic* (New York, 1971), 663–66.

14. Dickens, *Lollards and Protestants*, 48–49.

15. Hill, *World Turned Upside Down*, 21–31.

16. Buchanan Sharp, *In Contempt of All Authority: Rural Artisans and Riot in the West of England, 1586–1660* (Berkeley, 1980).

17. George Unwin, *Industrial Organization in the Sixteenth and Seventeenth Centuries* (Oxford, 1904), 48–52.

18. *The Mournfull Cryes of Many Thousand Poor Tradesmen . . . Or, the Warning Tears of the Oppressed* (London, 1648), in William Haller and Godfrey Davies, eds., *The Leveller Tracts, 1647–1653* (New York, 1944), 127, 126.

19. The Levellers have had many historians. Among the best are Hill, *World Turned Upside Down*, ch. 7; G. E. Alymer, *The Levellers in the English Revolution* (London, 1975), Introduction; and H. N. Brailsford, *The Levellers and the English Revolution*, ed. C. Hill (London, 1961).

20. On the dissenting backgrounds of the Leveller leaders, see Michael R. Watts, *The Dissenters: From the Reformation to the French Revolution*, Volume 1 (Oxford, 1978), 118.

21. On Lilburne, see Hill, *World Turned Upside Down*, ch. 7; Manning, *English People*, chs. 9 and 10; and Pauline Gregg, *Free-Born John; A Biography of John Lilburne* (London, 1961).

22. For a discussion of the grievances of London artisans, see Manning, *English People*, ch. 9.

23. William Walwyn, *The Power of Love* (London, 1643), quoted in ibid., 278.

24. [William Walwyn], *The Compassionate Samaritane* (London, 1645), quoted in ibid., 281–82.

25. "Mournfull Cryes," in Haller and Davies, *Leveller Tracts*, 129.

26. Ibid., 127.

27. Ibid., 126.

28. Ibid., 127.

29. *An Agreement of the Free People of England* (London, 1649), in ibid., 318–28.

30. Manning, *English People*, chs. 9 and 10; C. B. Macpherson, *The Political Theory of Possessive Individualism: Hobbes to Locke* (Oxford, 1962), ch. 3.

31. Manning, *English People*, ch. 9.

32. Ibid., ch. 10.

33. E. P. Thompson, *The Making of the English Working Class* (New York, 1966), 23.

34. Quoted in Manning, *English People*, 281.

35. On the flowering of dissent in the civil-war period see Hill, *World Turned Upside Down*.

36. On the preservation of radical ideas by the religious movements of the civil-war years, see the contributions in J. F. McGregor and B. Reay, eds., *Radical Religion in the English Revolution* (Oxford, 1984). For the post-Restoration years, see Richard L. Greaves, *Deliver Us from Evil: The Radical Underground in Britain, 1660–1663* (New York, 1986), and idem, *Enemies Under His Feet: Radicals and Non-Conformists in Britain, 1664–1677* (Stanford, 1990).

37. The place of the Ranters in early Quaker culture is discussed in Hill, *World Turned Upside Down*, chs. 9 and 10; A. L. Morton, *The World of the Ranters* (London, 1970); and Melvin B. Endy, Jr., *William Penn and Early Quakerism* (Princeton, 1973).

38. Jacob Barthumley, *The Light and Dark Sides of God* (1650), quoted in Hill, *World Turned Upside Down*, 165.

39. Edward Hide, *A Wonder, Yet No Wonder* (1651), quoted in ibid.

40. The history of early Quaker beliefs can be found in William C. Braithwaite, *The Beginnings of Quakerism* (London, 1923); Hugh Barbour, *The Quakers in Puritan England* (New Haven, Conn., 1964); and Hill, *World Turned Upside Down*, chs. 9 and 10.

41. William Penn, "Preface," to George Fox, *Journal*, 2 vols. (Philadelphia, 1839), I, xxviii.

42. Gary B. Nash, *Quakers and Politics: Pennsylvania, 1681–1726* (Princeton, 1968), ch. 1.

43. William T. Hull, *William Penn, a Topical Biography* (London, 1937), quoted in ibid., 43.

44. The provisions of the Frame of 1682 are discussed in ibid., 39–47.

45. Penn's early factional disputes are discussed in detail in ibid.

46. On artisan political participation in colonial Pennsylvania, see J. R. Pole, *Political Representation in England and the Origins of the American Republic* (Berkeley, 1966), 250–80; David Freeman Hawke, *In the Midst of a Revolution* (Philadelphia, 1961); and Gary B. Nash, *The Urban Crucible: Social Change, Political Consciousness, and the Origins of the American Revolution* (Cambridge, Mass., 1979).

47. Nash, *Quakers and Politics*, chs. 5 and 6.

48. Ibid., 114–27.

49. Among Penn's most important concessions were the transfer of effective power from the proprietor's council to the elected assembly, assembly appointment of country court positions, and the granting of a new city charter which made Philadelphia an independent entity. Ibid., 208–32.

50. The best account of the Lloyd-Logan dispute can be found in ibid., 332–35. On Lloyd, see Roy N. Lokken, *David Lloyd, Colonial Lawmaker* (Seattle, 1959).

51. Nash, *Urban Crucible*, 99–100.

52. Ibid., 306–18, 327–30.

53. The classic discussion of the decline of Quaker principles among Philadelphia merchants is Frederick B. Tolles, *Meeting House and Counting House: The Quaker Merchants of Colonial Philadelphia, 1682–1763* (Chapel Hill, 1948).

54. Nash, *Urban Crucible*, 103. The volume of immigration was substantial, reaching a total of 41,000 immigrants (45% German/55% Irish and Scots-Irish) between 1720 and 1739. See James G. Lydon, "Philadelphia's Commercial Expansion, 1720–1739," *Pennsylvania Magazine of History and Biography* 91 (1967), 407–8; Marianne S. Wokeck, "German and Irish Immigration to Colonial Philadelphia," *Proceedings of the American Philo-*

sophical Society 133 (1989), 128–43; and idem, "The Flow and the Composition of German Immigration to Philadelphia, 1727–1775," *Pennsylvania Magazine of History and Biography* 105 (1981), 249–78.

55. On the religious overtones of Philadelphia politics before the Revolution, see Dietmar Rothermund, *The Layman's Progress: Religious and Political Experience in Colonial Pennsylvania, 1740–1770* (Philadelphia, 1961); James H. Hutson, *Pennsylvania Politics, 1746–1770; The Movement for Royal Government and Its Consequences* (Princeton, 1972); and Sally Schwartz, *"A Mixed Multitude": The Struggle for Toleration in Colonial Pennsylvania* (New York, 1987), chs. 3–7.

56. James Logan, *The Charge Delivered from the Bench to the Grand Jury* (Philadelphia, 1723). Logan continued in this vein in A *Dialogue Showing What's Therein To Be Found* (Philadelphia, 1725).

57. Nash, *Quakers and Politics*, 320.

58. Ibid., 332.

59. Nash, *Urban Crucible*, 460, n. 81.

60. Logan to Henry Goldney, April 9, 1723, in Samuel Hazard et al., eds., *Pennsylvania Archives* (Harrisburg and Philadelphia, 1852–53), 2nd, ser., VII, 75.

61. William Keith, *The Observator's Trip to America, in a Dialogue between the Observator and His Country-man Roger* ([Philadelphia], 1726).

62. On the Leather Apron Club, see Thomas Wendel, "The Keith-Lloyd Alliance: Factional and Coalition Politics in Colonial Pennsylvania," *Pennsylvania Magazine of History and Biography* 92 (1968), 299–300.

63. On Keith's life, see Charles P. Keith, "Sir William Keith," *Pennsylvania Magazine of History and Biography* 12 (1888), 1–33, and Thomas H. Wendel, "The Life and Writings of Sir William Keith, Lieutenant-Governor of Pennsylvania and the Three Lower Counties, 1717–1726" (Ph.D. diss. University of Washington, 1964).

64. On Jacobite agitation among the small producers of England, see Linda Colley, "Eighteenth-Century English Radicalism Before Wilkes," *Transactions of the Royal Historical Society* 5th ser., 31 (1981), 1–19; R. J. Goulden, "Vox Populi, Vox Dei: Charles Delafaye's Paperchase," *The Book Collector* 28 (1979), 368–90; and Paul Kléber Monod, *Jacobitism and the English People, 1688–1788* (New York, 1989).

65. *Votes and Proceedings of the House of Representatives of the Province of Pennsylvania*, in Gertrude MacKinney, ed., *Pennsylvania Archives* (Harrisburg, Pa., 1931–35), 8th ser., II, 1459. For such ideas among the Levellers see the statements of William Dell, quoted in Hill, *World Turned Upside Down*, 80; Richard Overton, in ibid., 31; and Gerrard Winstanley, in G. H. Sabine, ed., *The Works of Gerrard Winstanley* (Ithaca, 1941), 253–54, 262.

66. *Votes and Proceedings*, II, 1460.

67. The club's street activities are mentioned in Wendel, "The Keith-Lloyd Alliance," 302. For an idea of club discussions, see the dialogue between Roger and the Observator in Keith, *Observator's Trip*.

68. David Lloyd, A *Vindication of the Legislative Power* (Philadelphia, 1725). The "Learned man" of "ill Temper" was James Logan.

69. A. S. P. Woodhouse, ed., *Puritanism and Liberty: Being the Army Debates* (1647–9) *from the Clarke Manuscripts with Supplementary Documents* (London, 1938, rpt., 1950), 59–60.

70. James Logan, *The Antidote: In Some Remarks on a Paper of David Lloyd's called "A Vindication of the Legislative Power"* (Philadelphia, 1725).

71. Ibid.

72. Ibid.

73. [Anon.], *The Triumvirate of Pennsylvania* (Philadelphia, 1729).

74. Ibid.

75. Benjamin Franklin, A *Modest Enquiry into the Nature and Necessity of a Paper-Currency*, in Leonard W. Labaree et al., eds., *The Papers of Benjamin Franklin* (New Haven, Conn., 1959–), I, 139–57. Franklin discussed his ideas on paper money with fellow craftsmen in his famous Junto for several months preceding the publication of his tract and found them "well received by the common people in general." See Carl Van Doren, *Benjamin Franklin* (New York, 1938), 101–2. The quote appears on p. 102.

76. Labaree et al., *Papers of Franklin*, 149.

77. Ibid., 150.

78. William Keith, A *Modest Reply to the Speech of Isaac Norris . . .* (Philadelphia, 1727).

79. Ibid.

80. The quote is from Keith, *Observator's Trip*.

81. On the transition from pre-political to political activity see Eric Hobsbawm, *Primitive Rebels: Studies in Archaic Forms of Social Movement in the 19th and 20th Centuries* (New York, 1959), and idem, *Bandits* (New York, 1981), 17–29.

82. Van Doren, *Franklin*, 74–80.

83. Ibid., 78.

84. On Philadelphia politics during the 1730s, see Alan Tully, *William Penn's Legacy: Politics and Social Structure in Provincial Pennsylvania, 1726–1755* (Baltimore, 1977), 3–22.

85. *Pennyslvania Gazette* (Philadelphia), March 25, 1736.

86. Ibid., April 1, 1736.

87. Ibid.

88. For the Putney Debates, see Woodhouse, *Puritanism and Liberty*.

89. *Pennsylvania Gazette*, April 8, 1736.

90. Ibid., May 6, 1736.

91. Ibid.

92. For Franklin's knowledge of the English tradition of popular militarism in defense of the common good, see his essay *Plain Truth*, in Labaree et al., *Papers of Franklin*, III, 188–204. See especially pp. 202–3 where he writes "BRITONS tho' a Hundred Years transplanted . . . may yet retain, even to the third and fourth Descent, that *Zeal* for the *Publick Good*, that *military Prowess*, and that *undaunted Spirit*, which has in every Age distinguished their Nation." That Franklin was well informed about the civil-war period during the organization of the Association can be inferred from his remembrances of Oliver and Richard Cromwell in *Poor Richard* for 1748 and 1750, and of the New Model Army general, Robert Black, in the 1749 edition. References to Cromwell and the civil war do not appear before these dates, thus it can be assumed that Franklin was reading about the civil war, and especially about its military aspects, for precedents and guidance in forming the Association. Interestingly, Franklin compares Oliver Cromwell to Caesar and suggests that had he come to the colonies, he would have risen to the rank of governor.

93. The history of Franklin's militia efforts can be followed in Van Doren, *Franklin*, 184–87.

94. The impact of the Seven Years' War is discussed in Nash, *Urban Crucible*, ch. 9.

95. Billy G. Smith, "The Material Lives of Laboring Philadelphians, 1750–1800," *William and Mary Quarterly* 3d ser., 38 (1981), 163–202.

96. Gary B. Nash, "Poverty and Poor Relief in Pre-Revolutionary Philadelphia," *William and Mary Quarterly* 3d ser., 33 (1976), 3–30, and idem, "Up from the Bottom in Franklin's Philadelphia," *Past and Present* 77 (1977), 57–83.

97. *To Several Battalions of Military Associators in the Province of Pennsylvania* (Philadelphia, 1776).

98. The story of Franklin's attempt to negotiate a royal charter is told in William S. Hanna, *Benjamin Franklin and Pennsylvania Politics* (Stanford, 1964), and Hutson, *Pennsylvania Politics*.

99. The incident is recounted in Benjamin H. Newcomb, "The Stamp Act and Pennsylvania Politics," *William and Mary Quarterly* 3d ser., 23 (1966); James H. Hutson, "An Investigation of the Inarticulate: Philadelphia's White Oaks," *William and Mary Quarterly* 3d ser., 28 (1971); and Jesse Lemisch and John K. Alexander, "The White Oaks, Jack Tar, and the Concept of the Inarticulate," *William and Mary Quarterly* 3d ser., 29 (1972), 109–36.

100. *To the Free and Patriotic Inhabitants of the City of Phila[delphia]* (Philadelphia, 1770).

101. George H. Sabine, ed., *The Works of Gerrard Winstanley* (New York, 1941, rpt., 1965), 181.

102. Hill, *World Turned Upside Down*, 31.

103. See p. 22, above.

104. Nash, *Urban Crucible*, 376–77.

105. Ibid., 378.

106. R. A. Ryerson, "Political Mobilization and the American Revolution: The Resistance Movement in Philadelphia, 1765 to 1776," *William and Mary Quarterly* 3d ser., 31 (1974), Fig. 2, p. 587; idem, *The Revolution Is Now Begun: The Radical Committees of Philadelphia, 1765–1776* (Philadelphia, 1978), 184–85, 190.

107. Steven Rosswurm, *Arms, Country, and Class: The Philadelphia Militia and the "Lower Sort" during the American Revolution* (New Brunswick, N.J., 1987), Table A.2, 260–61.

108. Ibid., 156.

109. Ibid., 121.

110. Ibid., 150.

111. "Memorial of the Committee of Privates," May 11, 1776, quoted in ibid., 158.

112. *Pennsylvania Gazette*, March 13, 1776.

113. Ibid.

CHAPTER 2
PROPERTY RIGHTS AND COMMUNITY RIGHTS

1. *Pennsylvania Evening Post* (Philadelphia), April 30, 1776.

2. Gary B. Nash, *The Urban Crucible: Social Change, Political Consciousness, and the Origins of the American Revolution* (Cambridge, Mass., 1979), Table 13, pp. 407–8; Susan E. Klepp, "Demography in Early Philadelphia, 1690–1860," *Proceedings of the American Philosophical Society* 133 (1989), Table 2, p. 104; P. M. G. Harris, "The Demographic Development of Colonial Philadelphia in Some Comparative Perspective," *Proceedings of the American Philosophical Society* 133 (1989), Table 1, p. 274.

3. Klepp, "Demography in Early Philadelphia," Table 2, p. 105.

4. Ibid., Fig. 2, pp. 410–11; James G. Lydon, "Philadelphia's Commercial Expansion, 1720–1739," *Pennsylvania Magazine of History and Biography* 91 (1967), 401–18; Marc Egnal, "The Changing Structure of Philadelphia's Trade with the British West Indies, 1750–1775," ibid., 99 (1975), 156–79.

5. Nash, *Urban Crucible*, Fig. 2, pp. 410–11.

6. Billy G. Smith, "The Material Lives of Laboring Philadelphians, 1750–

1800," *William and Mary Quarterly* 3d ser., 38 (1981), 183–200; idem, *The "Lower Sort": Philadelphia's Laboring People, 1750–1800* (Ithaca, 1990), ch. 3; Nash, *Urban Crucible*, 238–44, 392–94, 397–98; idem, "Up from the Bottom in Franklin's Philadelphia," *Past and Present* 77 (1977), 70–71.

7. Smith, "Material Lives," Table II, p. 173; Nash, *Urban Crucible*, 322.

8. Nash, *Urban Crucible*, 322.

9. Ibid., Table 4, p. 396.

10. Ibid.

11. Ibid.

12. Ibid.

13. Thomas M. Doerflinger, *A Vigorous Spirit of Enterprise: Merchants and Economic Development in Revolutionary Philadelphia* (Chapel Hill, 1986), chs. 1–3.

14. On laboring-class support for the Levellers, see Brian Manning, *The English People and the English Revolution, 1640–1649* (London, 1976). On artisans in the French Revolution, see Albert Soboul, *The Sans-Culottes: The Popular Movement and Revolutionary Government, 1793–1794* (Princeton, 1980).

15. The best discussions of Philadelphia's bound laborers are Sharon V. Salinger, "Artisans, Journeymen, and the Transformation of Labor in Late Eighteenth-Century Philadelphia," *William and Mary Quarterly* 3d ser., 40 (1983), 62–84; and idem, *"To Serve Well and Faithfully": Labor and Indentured Servants in Pennsylvania, 1682–1800* (New York, 1987). The general question of bound labor in colonial America is addressed in Richard S. Dunn, "Servants and Slaves: The Recruitment and Employment of Labor," in Jack P. Greene and J. R. Pole, eds., *Colonial British America: Essays in the New History of the Early Modern Era* (Baltimore, 1984), 157–94.

16. Salinger, *To Serve Well and Faithfully*, Table A3, pp. 178–80; idem, "Transformation of Labor," 64–66.

17. Billy G. Smith, "The Vicissitudes of Fortune: The Careers of Laboring Men in Philadelphia, 1750–1800," in Stephen Innes, ed., *Work and Labor in Early America* (Chapel Hill, 1988), 221–51.

18. Ian M. G. Quimby, "Apprenticeship in Colonial Philadelphia" (M.A. thesis, University of Delaware, 1963).

19. Salinger, "Transformation of Labor," 74.

20. Ibid., 72–74.

21. See Chapter 5, below.

22. The making of the Constitution of 1776 is discussed in J. Paul Selsam, *The Pennsylvania Constitution of 1776: A Study in Revolutionary Democracy* (Philadelphia, 1936), and Theodore Thayer, *Pennsylvania Politics and the Growth of Democracy, 1740–1776* (Harrisburg, Pa., 1953), ch. 7.

23. Thayer, *Pennsylvania Politics*, 184.

24. On the Agitator's demands, see Christopher Hill, *The World Turned Upside Down: Radical Ideas During the English Revolution* (New York, 1972), ch. 7.

25. *The Constitution of the Common-wealth of Pennsylvania* (Philadelphia, 1776).

26. On constitution-making in the states, see Willi Paul Adams, *The First American Constitutions: Republican Ideology and the Making of the State Constitutions in the Revolutionary Era* (Chapel Hill, 1980); Gordon S. Wood, *The Creation of the American Republic, 1776–1787* (Chapel Hill, 1969), chs. 4–6; and J. R. Pole, *Political Representation in England and the Origins of the American Republic* (London, 1966).

27. *Pennsylvania Gazette* (Philadelphia), June 11, 1777.

28. *Constitution of the Common-wealth.*

29. *An Essay of a Declaration of Rights* (Philadelphia, 1776).

30. Ibid. My italics.

31. Ibid.

32. On the relationship between the English and American revolutions, see John M. Murrin, "The Great Inversion, or Court versus Country: A Comparison of the Revolution Settlements in England (1688–1721) and America (1776–1816)," in J. G. A. Pocock, ed., *Three British Revolutions: 1641, 1688, 1776* (Princeton, 1980), 368–453.

33. Alexander Graydon, *Memoirs* (Philadelphia, 1846), 283–84.

34. Robert Proud, "On the Violation of Established and Lawful Order," quoted in Selsam, *Pennsylvania Constitution*, 209.

35. "Orator Puff," *Pennsylvania Evening Post* (Philadelphia), Oct. 9, 1776.

36. The history of Philadelphia's Revolutionary militia is told in Steven Rosswurm, *Arms, Country, and Class: The Philadelphia Militia and the "Lower Sort" During the American Revolution* (New Brunswick, N.J., 1987).

37. Ibid., ch. 4.

38. The Associators' role in the winter campaign of 1776–77 is described in Rosswurm, *Arms, Country, and Class*, 123–35.

39. *Pennsylvania Gazette* (Philadelphia), June 18, 1777.

40. William Williams to Jonathan Trumbull, Philadelphia, July 5, 1777, in Edmund B. Burnett, ed., *Letters of Members of the Continental Congress*, 8 vols. (Washington, D.C., 1921–36), II, 401.

41. Samuel Hazard et al., eds., *Pennsylvania Archives* (Harrisburg and Philadelphia, 1852–53), Ser. 1, VII, 392.

42. Ibid., 393, 394.

43. Ibid., 393.

44. *Pennsylvania Packet* (Philadelphia), June 3, 1777.

45. *Pennsylvania Archives*, Ser. 1, VII, 394.

46. The British occupation is discussed in John W. Jackson, *With the British Army in Philadelphia, 1777–1778* (San Rafael, Calif., 1979).

47. The "Meschianza" is discussed most fully in the *Royal Pennsylvania Gazette* (Philadelphia), May 26, 1778. Other accounts appear in J. Thomas Scharf and Thompson Westcott, *A History of Philadelphia, 1609–1884*, 3 vols. (Philadelphia, 1884), I, 378–82; Ellis P. Oberholtzer, *Philadelphia: A History of the City and Its People*, 4 vols. (Philadelphia, 1912), I, 274–76; and John F. Watson, *Annals of Philadelphia, and Pennsylvania, in the Olden Time*, 3 vols. (Philadelphia, 1900), II, 290–93.

48. Scharf and Westcott, *History of Philadelphia*, I, 381.

49. Rosswurm, *Arms, Country, and Class*, 152.

50. Ibid., 154–55.

51. Ibid. A somewhat different assessment appears in his "Arms, Culture, and Class: The Philadelphia Militia and the 'Lower Orders' in the American Revolution, 1765–1783" (Ph.D. diss., Northern Illinois University, 1979), 305–6.

52. Pennsylvania Assembly, "An Act for Regulating the Prices," April 1, 1778, in James T. Mitchell and Henry Flanders, comps., *The Statutes at Large of Pennsylvania from 1682 to 1801*, 17 vols. (Harrisburg, Pa., 1896–1915), IX, 236–38.

53. *Pennsylvania Packet*, Dec. 10, 1778.

54. *Pennsylvania Archives*, Ser. 1, VII, 393.

55. Ibid., 392.

56. Ibid.

57. Samuel Rowland Fisher to Jabez Maud Fisher, Feb. 19, 1779, quoted in Rosswurm, "Arms, Culture, and Class," 359.

58. Robert L. Brunhouse, *The Counter-revolution in Pennsylvania, 1776–1790* (Harrisburg, Pa., 1942), ch. 1 and passim.

59. Graydon, *Memoirs*, 285.

60. Ibid., 288.

61. Benjamin Rush, *The Letters of Benjamin Rush*, ed. L. H. Butterfield, 2 vols. (Princeton, 1951), I, 440.

62. For the social composition of the Republican Society, see Rosswurm, *Arms, Country, and Class*, 176–77.

63. See ibid., 176 for the leaders of the Constitutional Society; and David Freeman Hawke, *In the Midst of a Revolution* (Philadelphia, 1961), 102–7 for the leaders of the independence movement in Philadelphia.

64. On Harbeson and the "steering committee," see Hawke, *Midst of a Revolution*, 130–31; on Schlosser, see Richard Alan Ryerson, *The Revolution Is Now Begun: The Radical Committees of Philadelphia, 1765–1776* (Philadelphia, 1978), Appendix M, 280.

65. Rosswurm, *Arms, Country, and Class*, 177–78.

66. Quoted in ibid., 178.

67. Ibid.

68. William T. Read, *The Life and Correspondence of James Read* (Philadelphia, 1870), 349; John Wallace, *An Old Philadelphian. Colonel William Bradford* (Philadelphia, 1884), 305.

69. *Pennsylvania Packet*, May 27, 1779. Joseph Stansberry, a Tory china merchant and amateur poet, described the meeting in satiric verse—see Watson, *Annals*, II, 304–5.

70. Eric Foner, *Tom Paine and Revolutionary America* (New York, 1976), 168.

71. The activities of the Committee of Trade can be followed in ibid., ch. 5; and Rosswurm, *Arms, Country, and Class*, 179–94.

72. City Committee to Joseph Reed, June 5, 1779, quoted in Rosswurm, *Arms, Country, and Class*, 185–86.

73. *Pennsylvania Packet*, June 12, 1779.

74. Elaine Forman Crane et al., eds., *The Diary of Elizabeth Drinker*, 3 vols. (Boston, 1991), I, 355.

75. On the citizen's plan, see Foner, *Paine*, 173–74.

76. The citizen's plan supported the merchant claim that depreciating Revolutionary currency was the true cause of the city's rampant inflation. The only significant difference was that the citizen's plan rested upon voluntary subscriptions rather than enforced taxation.

77. The quote is from Franklin, *Plain Truth*, in Leonard W. Labaree et al., eds., *The Papers of Benjamin Franklin* (New Haven, Conn., 1959–   ), III, 188–204.

78. Rosswurm, *Arms, Country, and Class*, 176; on Matlack see Foner, *Paine*, 109–11.

79. On Paine, see Foner, *Paine*, ch. 6; on Marshall and Rush, see Hawke, *Midst of a Revolution*, 177–78; on Young and Cannon, see Rosswurm, "Arms, Culture, and Class," Appendix D, 501–3; and David Freeman Hawke, "Dr. Thomas Young—'Eternal Fisher in Troubled Waters': Notes for a Biography," *New York Historical Society Quarterly* 54 (1970), 7–29. On Cannon's authorship of the Pennsylvania Constitution of 1776, see Ryerson, *Revolution Has Begun*, 241, n. 147.

80. See the Constitutional Society announcement of a subscription to be taken in each ward for the "support of the necessitous Families of those unfortunate Men who were killed and wounded [at Ft. Wilson]," *Pennsylvania Packet*, Oct. 9, 1779. The leaders' names appear at the end of the announcement, nearly all bearing the rank of captain.

81. William B. Reed, *The Life and Correspondence of Joseph Reed* (Philadelphia, 1847), quoted in Rosswurm, "Arms, Culture, and Class," 445.

82. Crane, ed., *Drinker Diary*, I, 361.

83. The events at Fort Wilson are recounted in John K. Alexander, "The Fort Wilson Incident of 1779: A Case Study of the Revolutionary Crowd," *William and Mary Quarterly* 3d ser., 31 (1974), 589–612.

84. Joseph Reed, "Proclamation," May 11, 1781, quoted in Rosswurm, *Arms, Country, and Class*, 244.

85. Ibid., 243.

86. Smith, "Material Lives," Table II, p. 173. The household budget includes only the basic necessities of food, rent, firewood, and clothing.

87. George Nelson Diary, April 17, 1780, quoted in Rosswurm, *Arms, Country, and Class*, 232.

88. Ibid., 245.

89. Ibid., 243.

90. Ibid., 240–41.

91. Foner, *Paine*, 188.

92. The now-classic text is E. P. Thompson, "The Moral Economy of the English Crowd in the Eighteenth Century," *Past and Present* 50 (1971), 76–136.

93. Ibid., 126–31; idem, "Eighteenth-Century English Society: Class Struggle Without Class?," *Social History* 3 (1978), 133–65.

94. The debate appeared in the *Pennsylvania Packet*, Sept. 10, 1779. The full text of the debate is reprinted in my "Small Producer Thought in Early America, Part I: Philadelphia Artisans and Price Controls," *Pennsylvania History* 54 (1987), 115–47.

95. *Pennsylvania Packet*, Sept. 10, 1779.

96. Ibid.

97. The members of the Committee for Enquiring into the State of Trade (the Committee of Thirteen) are listed in *Pennsylvania Packet*, Aug. 12, 1779. The quote is from ibid., Sept. 10, 1779.

98. *Pennsylvania Packet*, Sept. 10, 1779.

99. Ibid.

100. Ibid.

101. Ibid.

102. Ibid.

103. Ibid.

104. In *The Law of Freedom in a Platform or True Magistracy Restored* (1652), the Digger Leader Gerrard Winstanley argued that

> No man can be rich, but he must be rich, either by his own labors, or by the labors of other men helping him: If a man have no help from his neighbor, he shall never gather an Estate of hundreds and thousands a year: If other men help him to work, then are those Riches his Neighbors, as well as his; for they be the fruit of other men's labors as well as his own.

The Works of Gerrard Winstanley, George H. Sabine, ed. (New York, 1941, rpt., 1965), 511.

CHAPTER 3
REACTION AND RESTORATION

1. The events immediately following Fort Wilson are recounted in Steven Rosswurm, *Arms, Country, and Class: The Philadelphia Militia and the "Lower Sort" During the American Revolution* (New Brunswick, N.J., 1987), 217–22.

2. Samuel Hazard et al., eds., *Pennsylvania Archives* (Harrisburg and Philadelphia, 1852–53), 1st. ser., VII, 732.

3. Benjamin Rush, "Address to the People of the United States," *American Museum* 1 (1787), 9.

4. *Pennsylvania Packet* (Philadelphia), Jan. 27, 1784.

5. The most thorough account of Pennsylvania politics in the years between the Revolution and 1790 is Robert L. Brunhouse, *The Counter-revolution in Pennsylvania, 1776–1790* (Harrisburg, Pa., 1942). Another account, drawing upon different sources, is E. Bruce Thomas, *Political Tendencies in Pennsylvania, 1783–1794* (Philadelphia, 1938).

6. There was an apparent agreement between the two parties to keep the issue of the constitution out of the political arena until the censorial election in 1783. See Samuel B. Harding, "Party Struggles over the First Pennsylvania Constitution," *Annual Report of the American Historical Association for the Year 1894* (Washington, D.C., 1895), 385, n. 2.

7. The majority report appeared in the *Journal of the Council of Censors*, Jan. 19, 1784. Cf. *An Address of the Council of Censors to the Freemen of Pennsylvania* (Philadelphia, 1784).

8. The minority address was printed in the *Pennsylvania Packet*, Jan. 27, 1784; and separately as Council of Censors, *To the Freemen of Pennsylvania* (Philadelphia, 1784).

9. Ibid.

10. Ibid.

11. For one such remonstrance, see the *Pennsylvania Packet*, Feb. 12, 1784.

12. The Republican disclaimer is quoted in Thomas, *Political Tendencies*, 75.

13. The 1790 constitution is reprinted in John H. Fertig and Frank M. Hunter, comps., *Constitutions of Pennsylvania* (Harrisburg, Pa., 1916), 189–211.

14. Brunhouse, *Counter-revolution*, 152.

15. *Pennsylvania Packet*, June 4, 1785.

16. Ibid., June 17, 1785.

17. *Freeman's Journal* (Philadelphia), June 1, 1785.

18. *Pennsylvania Packet*, June 18, 1785.

19. Ibid.

20. George Bryan, "An Address of the Subscribers . . . ," *Pennsylvania Packet*, Oct. 4, 1787.

21. Ibid., 13.

22. Hutchinson's essays appeared in the *Independent Gazetteer* between Oct. 12, 1787, and Feb. 6, 1788. They are reprinted in Herbert J. Storing, ed., *The Complete Anti-Federalist*, 7 vols. (Chicago, 1981), III, 17–52.

23. Storing, *Anti-Federalist*, III, 18–19. The general sentiment in the city is best captured by "Philadelphiensis," who wrote, "Were some additional powers for regulating commerce, and the *impost duties* for a limited time, granted to the present Congress, this would probably answer for all our purposes." Ibid., 112.

24. Storing, *Anti-Federalist*, III, 19.

25. Ibid., 19.

26. Ibid.

27. Ibid., 23. See "Old Whig," essays II and V for similar arguments, ibid., 22–26 and 34–38.

28. Ibid., 23–24.

29. Ibid., 39.

30. *Independent Gazetteer* (Philadelphia), Oct. 17, 1787. Cf. the essays by "Philadelphiensis," in Storing, *Anti-Federalist*, III, 99–140, for similar arguments in more formal terms.

31. Ibid., 53.

32. Ibid., 54.

33. Ibid.

34. *Independent Gazetteer*, Oct. 29, 1787.

35. Storing, *Anti-Federalist*, III, 89. On popular protest in England and Philadelphia, see Chapter 1, above. For an argument that links late eighteenth-century Anti-Federalism and nineteenth-century democratic politics, see Gordon S. Wood, "Interests and Disinterestedness in the Making of the Constitution," in Richard Beeman, Stephen Botein, and Edward C. Carter II, eds., *Beyond Confederation: Origins of the Constitution and American National Identity* (Chapel Hill, 1987), 69–109.

36. Storing, *Anti-Federalist*, III, 89.

37. Ibid., 89–90.

38. Ibid., 90.

39. *Pennsylvania Gazette*, Oct. 3, 1787, reprinted in Merrill Jensen, ed.,

The Documentary History of the Ratification of the Constitution, 16 vols. (Madison, Wis., 1976–   ), II, 157.

40. *Pennsylvania Gazette*, Oct. 17, 1787, in Jensen, *Documentary History*, II, 186.

41. Ibid., 187.

42. Ibid., 188.

43. Ibid.

44. In addition to the above, see Benjamin Rush's speech to the town meeting held on Oct. 6, 1787, in Jensen, *Documentary History*, II, 174–75.

45. The meeting was held on the same day that the text of the Constitution was made public. For the petition, see Jensen, *Documentary History*, II, 134. For reports of other meetings and petitions to the Pennsylvania legislature, see ibid., 64–65, 130, 134, 167–72, 174–75, 724.

46. David Redick to William Irvine, Philadelphia, Sept. 24, 1787, in ibid., 135.

47. Charles Olton, *Artisans for Independence: Philadelphia Mechanics and the American Revolution* (Syracuse, N.Y., 1975), 119; Brunhouse, *Counter-revolution*, 207. Alfred Young, *The Democratic Republicans of New York: The Origins, 1763–1797* (Chapel Hill, 1967), 569–70, and Howard Rock, *Artisans of the New Republic: The Tradesmen of New York City in the Age of Jefferson* (New York, 1984), 22–23, make a similar case for New York artisans, as does Charles G. Steffen, *The Mechanics of Baltimore: Workers and Politics in the Age of Revolution, 1763–1812* (Urbana, Ill., 1984), ch. 4, for the artisans of Baltimore.

48. For modern historians, see note 47. For a contemporary assessment, see George Bryan's remarks quoted in Olton, *Artisans for Independence*, 114.

49. Brunhouse, *Counter-revolution*, 206–7.

50. Jackson Turner Main, *The Antifederalists: Critics of the Constitution, 1781–88* (New York, 1961), 252–59.

51. Brunhouse, *Counter-revolution*, 172–73.

52. *Independent Gazetteer*, April 23, 1788.

53. Ibid.

54. Ibid.

55. Ibid.

56. Ibid.

57. Ibid.

58. George Bryan, "Account of the Adoption of the Constitution," quoted in Olton, *Artisans for Independence*, 114.

59. *Independent Gazetteer*, Feb. 15, 1788.

60. The history of the bank wars of the 1780s can be followed in Joseph Dorfman, *The Economic Mind in American Civilization, 1606–1865*, 2 vols.

(New York, 1946), I, 221–27; and Clarence L. Ver Steeg, *Robert Morris: Revolutionary Financier* (Philadelphia, 1954), chs. 4–8.

61. E. Hazard to Belknap, Jan. 24, 1784, Belknap Papers, quoted in Brunhouse, *Counter-revolution*, 150.

62. Mary M. Schweitzer, *Custom and Contract: Household, Government, and the Economy in Colonial Pennsylvania* (New York, 1987), 7–8.

63. Ibid., Table 5.7, pp. 157–59.

64. Smilie in the *Pennsylvania Evening Herald* (Philadelphia), April 5, 1786, quoted in Main, *Anti-Federalists*, 187.

65. *Independent Gazetteer*, Feb. 28, 1787.

66. *Pennsylvania Gazette* (Philadelphia), May 30, 1785.

67. Ibid.

68. Ibid.

69. Ibid.

70. *Pennsylvania Packet*, March 25, 1786.

71. For the colonial Corporation, see Judith M. Diamondstone, "Philadelphia's Municipal Corporation, 1701–1776," *Pennsylvania Magazine of History and Biography* 90 (1966), 183–201.

72. Ibid., 194.

73. Journal of William Black, 1744, quoted in ibid., 193.

74. George Winthrop Geib, "A History of Philadelphia, 1776–1789" (Ph.D. diss., University of Wisconsin, 1969), 286.

75. The provisions of the city charter of 1789 are discussed in John K. Alexander, *Render Them Submissive: Responses to Poverty in Philadelphia, 1760–1800* (Amherst, Mass., 1980), 44–47.

76. Geib, "History of Philadelphia," 294.

77. Samuel Coates, "Cases of Several Lunatics in the Pennsylvania Hospital, and the Causes There Of in Many Cases," *Pennsylvania Magazine of History and Biography* 106 (1982), 283–86.

78. While no work has focused specifically on the political allegiances of laboring-class Philadelphians in the postwar era, this assumption informs Brunhouse, *Counter-revolution*, and Thomas, *Political Tendencies*.

79. The quote is from Thomas Paine, "To the Citizens of Pennsylvania," in Philip S. Foner, *The Complete Writings of Thomas Paine*, 2 vols. (New York, 1945), II, 993.

80. On the lack of residential segregation, see Sam Bass Warner, *The Private City: Philadelphia in Three Periods of Its Growth* (Philadelphia, 1968), 10–11. This is confirmed by an inspection of the 1782 tax assessments which show clearly that rich and poor were found in every ward. The 1782 assessments are in Samuel Hazard et al., eds., *Pennsylvania Archives* (Harrisburg and Philadelphia, 1852–53), 3d ser., XVI, 315–515.

81. This assumption runs through "A Philadelphia Mechanic," *Independent Gazetteer*, Oct. 8, 1785, and "A Friend to Mechanics," ibid., Oct. 11, 1785.

82. The classic account of English working-class formation is E. P. Thompson, *The Making of the English Working Class* (New York, 1966).

83. The broadside is reprinted in Rosswurm, *Arms, Country, and Class*, 231–32.

84. *Pennsylvania Gazette*, June 25, 1783.

85. Ibid.

86. Ibid., July 12, 1783.

87. Ibid., Oct. 11, 1783.

88. Ibid.

89. Ibid.

90. On the early organizations of English textile workers, see H. A. Turner, *Trade Union Growth, Structure and Policy: A Comparative Study of the Cotton Unions* (London, 1967), 44–78. On the informal and formal organizations of workers in eighteenth-century England, see John Rule, *The Experience of Labour in Eighteenth-Century Industry* (London, 1981), chs. 6–7; idem, *The Labouring Classes in Early Industrial England, 1750–1850* (London, 1986), ch. 11; C. R. Dobson, *Masters and Journeymen. A Prehistory of Industrial Relations, 1717–1800* (London, 1980), Appendix, 154–70; Robert W. Malcolmson, "Workers' Combinations in Eighteenth-Century England," in Margaret Jacob and James Jacob, eds., *The Origins of Anglo-American Radicalism* (London, 1984), 169–81; and the essays in John Rule, ed., *British Trade Unionism, 1750–1850: The Formative Years* (London, 1988). Similar organizations existed among the immigrant English mill workers of early nineteenth-century Philadelphia. See Cynthia J. Shelton, *The Mills of Manayunk: Industrialization and Social Conflict in the Philadelphia Region, 1787–1837* (Baltimore, 1986), ch. 2.

91. *Pennsylvania Gazette*, Oct. 11, 1783.

92. "The Mournful Cryes of Many Poor Tradesmen . . . ," in William Haller and Godfrey Davies, eds., *The Leveller Tracts, 1647–1653* (New York, 1944), 126.

93. *Mechanics' Free Press* (Philadelphia), Aug. 16, 1828.

94. *Independent Gazetteer*, Oct. 11, 1783.

95. Ibid.

96. Ibid.

97. Ibid.

98. Ibid.

99. *Pennsylvania Gazette*, Oct. 9, 1784.

100. Ibid.

101. Ibid.

102. *Freeman's Journal*, April 27, 1785.

103. For the identity of Joyce, Jr., see Alfred F. Young, "English Plebeian Culture and Eighteenth-Century American Radicalism," in Jacob and Jacob, *Anglo-American Radicalism*, 200. For Joyce's role in the English civil war, see Christopher Hill, *The Century of Revolution, 1603–1714* (New York, 1961), 130.

104. *Pennsylvania Gazette*, May 4, 1785.

105. Ibid.

106. Ibid.

107. The events of the town meeting were reported in, *Pennsylvania Gazette*, June 22, 1785; *Freeman's Journal*, June 22, 1785; *Pennsylvania Packet*, June 21, 1785; and *Pennsylvania Mercury* (Philadelphia), June 24, 1785.

108. The members of the Merchants' and Mechanics' committees are listed in J. Thomas Scharf and Thompson Westcott, *A History of Philadelphia, 1609–1884*, 3 vols. (Philadelphia, 1884), I, 439. The mean wealth of the two committees was calculated from the 1782 Philadelphia tax assessments in Hazard, *Pennsylvania Archives* 3d ser., XVI, 315–515. The mean wealth of the merchants was £1,456.

109. *Pennsylvania Gazette*, July 27, 1785.

110. Ibid., Sept. 14, 1785.

CHAPTER 4
HEGEMONY AND COUNTER-HEGEMONY

1. Anon., "A New Catichism: More Studied Than an Older and a Better One," *Independent Gazetteer* (Philadelphia), July 14, 1789.

2. Ibid., July 21, 1789.

3. Ibid.

4. Similar arguments in favor of the Constitution of 1776 can be found in ibid., Aug. 1 and 3, 1789.

5. Ibid., Aug. 12, 1789.

6. Alexander Hamilton et al., *The Federalist Papers* (New York, 1961), 77–78.

7. Ibid., 78.

8. Ibid., 79.

9. Ibid., 81.

10. Ibid., 82.

11. Pelatiah Webster, *Political Essays on the Nature and Operation of Money, Public Finances, and Other Subjects* (New York, 1791, rpt., 1968), 9.

12. Hamilton et al., *Federalist Papers*, 79.

13. Ibid., 78. This same argument provided the ideological foundation of the Whig party during the 1830s and 1840s. See John Ashworth, *"Agrarians"* & *"Aristocrats": Party Political Ideology in the United States, 1837–1846* (London, 1983).

14. Ibid., 84.

15. Ibid., 81.

16. *Independent Gazetteer*, April 9, 1791.

17. Ibid.

18. Ibid., May 30, 1795.

19. Richard G. Miller, "The Federal City, 1783–1800," in Russell F. Weigley, ed., *Philadelphia: A 300-Year History* (New York, 1982), 177.

20. Theodore Sedgwick to Ephraim Williams, June 1789, Sedgwick Papers, Massachusetts Historical Society, quoted in Joseph Charles, *The Origins of the American Party System: Three Essays* (New York, 1961), 38.

21. Webster, *Political Essays*, 215.

22. John Witherspoon, *Lectures on Moral Philosophy*, quoted in Joseph Dorfman, *The Economic Mind in American Civilization*, 2 vols. (New York, 1946), I, 126.

23. Webster, *Political Essays*, 215–16.

24. Ibid., 215.

25. Ibid., 216.

26. Ibid., 217.

27. Ibid.

28. The modern concept of hegemony has been the subject of a considerable literature. Drawing from the writings of Antonio Gramsci in the 1930s, hegemony became, in the 1970s and 1980s, an important concept with many conflicting definitions. The point here is that throughout the new nation men like Webster were attempting to consolidate the social and political position of merchants, speculators, and other men of wealth. An important part of their project was an attempt to transform popular support for the 1787 Constitution into an acceptance of mercantile and, in varying degrees, manufacturing capitalism. In Philadelphia this project became a contest between capitalists and small producers for the minds and allegiances of the city's laboring classes. The outcome of this contest, it is worth mentioning, was in no way preordained. The most useful works dealing with the concept of hegemony are Raymond Williams, *Marxism and Literature* (Oxford, 1977), ch. 6; Anne Showstack Sassoon, *Gramsci's Politics* (London, 1980); and Joseph V. Femia, *Gramsci's Political Thought: Hegemony, Consciousness, and the Revolutionary Process* (Oxford, 1987).

29. Webster, *Political Essays*, 71.

30. Ibid., 22.

31. Ibid., 218.

32. Ibid., 66.

33. Ibid., 9.

34. Ibid., 10.

35. Ibid., 24.

36. Ibid., 407.

37. Joel Barlow, *Advice to the Privileged Orders in the Several States of Europe* (Philadelphia, 1792, 1795).

38. Thomas Paine, *A Serious Address to the People of Pennsylvania on the Present Situation of Their Affairs* (Philadelphia, 1778), in Philip S. Foner, ed., *The Complete Writings of Thomas Paine*, 2 vols. (New York, 1945), II, 283.

39. James A. Bayard, *Annals of Congress*, 7th Cong., 2d sess., quoted in Drew R. McCoy, *The Elusive Republic: Political Economy in Jeffersonian America* (Chapel Hill, 1980), 179.

40. George Logan, *A Letter to the Citizens of Pennsylvania*, 2d ed. (Philadelphia, 1800).

41. On the political and economic choices of the 1790s, see McCoy, *Elusive Republic*; and Lance Banning, *The Jeffersonian Persuasion: Evolution of a Party Ideology* (Ithaca, 1978), for republican interpretations of the era. The liberal interpretation is advanced most cogently by Joyce Appleby, *Capitalism and a New Social Order: The Republican Vision of the 1790s* (New York, 1984). Also see Steven Watts, *The Republic Reborn: War and the Making of Liberal America, 1790–1820* (Baltimore, 1987). My point here is not to argue the merits of either position, but to point out that *both* sides ignore the role of indigenous popular moral traditions in the politics of the era.

42. Congressional debate on the Bankruptcy Act is discussed in McCoy, *Elusive Republic*, 178–84.

43. James A. Bayard and Harrison Gray Otis, *Annals of Congress*, 5th Cong., 3d sess., quoted in McCoy, *Elusive Republic*, 180.

44. Ibid., 181.

45. George W. Corner, ed., *The Autobiography of Benjamin Rush* (Philadelphia, 1948), 160.

46. Phineas Bond, "The Letters of Phineas Bond," *Annual Report of the American Historical Association for the Year 1896*, 2 vols. (1897), I, 568.

47. Ibid., 586–87.

48. Billy G. Smith, "The Material Lives of Laboring Philadelphians, 1750–1800," *William and Mary Quarterly* 3d ser., 38 (1981), Table III, 184, Table IV, 192, Table V, 195, Table VI, 199, and Fig. III and IV, 186.

49. George Winthrop Geib, "A History of Philadelphia, 1776–1784" (Ph.D. diss., University of Wisconsin, 1969), 203.

50. Ibid., 204–5. On the general rise in poverty and poor-relief costs in this period, see Billy G. Smith, "Poverty and Economic Marginality in Eighteenth-Century America," *Proceedings of the American Philosophical Society*, 32 (1988), 85–118.

51. *American Museum* II (Oct. 1788), 386.

52. Smith, "Material Lives," 202.

53. *Pennsylvania Gazette* (Philadelphia), June 12, 1791.

54. On the bubble and its effects, see Rush, *Autobiography*, 204, 217–18; and L. H. Butterfield, ed., *The Letters of Benjamin Rush*, 2 vols. (Princeton, 1951), I, 602–3.

55. Billy G. Smith, *The "Lower Sort": Philadelphia's Laboring People*, 1750–1800 (Ithaca, 1990), ch. 5.

56. Ibid., and idem, "Material Lives," 202.

57. John R. Commons et al., *History of Labour in the United States*, 2 vols. (New York, 1926), I, 108–37; J. Thomas Scharf and Thompson Westcott, *A History of Philadelphia*, 1609–1884, 3 vols. (Philadelphia, 1884), I, 465; *Independent Gazetteer*, June 9, 1792; The Journeymen Cabinet- and Chair-Makers, *Philadelphia Book of Prices*, 2d ed. (Philadelphia, 1795).

58. Robert W. Malcolmson, "Workers' Combinations in Eighteenth-Century England," in Margaret Jacob and James Jacob, eds., *The Origins of Anglo-American Radicalism* (London, 1984), 149–61; for a later period, see Dona Torr, *Tom Mann and His Times: Volume 1, 1856–1890* (London, 1956).

59. Henry P. Rosemont, "Benjamin Franklin and the Philadelphia Typographical Strikers of 1786," *Labor History* 22 (1981), 398–429; Ethelbert Stewart, "A Documentary History of the Early Organization of Printers," in U.S. Bureau of Labor, *Bulletin 61* (Washington, D.C., 1905), 857–1033.

60. Timothy Claxton, *Memoir of a Mechanic* (New York, 1839).

61. Rosemont, "Typographical Strikers," 406–7. The strike was successful, but the society fractured nonetheless, leaving the supporters of benevolence in control of the society while the trade unionists disappeared from public view. It is possible that the trade unionists played a role in creating the Philadelphia Typographical Society discussed on pages 120–23, below.

62. Journeymen Cabinet- and Chair-Makers, *Book of Prices*.

63. The price-setting function of the journeymen's societies is confirmed by the brief histories in Thomas Porter, *Picture of Philadelphia* (Philadelphia, 1831), 267–76.

64. On the colonial wage system, see Richard B. Morris, *Government and Labor in Early America* (New York, 1946), ch. 4.

65. Rosemont, "Typographical Strikers."

66. John R. Commons and Helen L. Sumner, eds., *A Documentary History of American Industrial Society*, 10 vols. (Cleveland, Ohio, 1910),

III, 60–248. This notion was still alive in 1828—see "Crispin," *Mechanics' Free Press* (Philadelphia), May 3, 1828.

67. The list is reprinted in Stewart, "Early Organizations of Printers," 865.

68. Ibid.

69. Ibid.

70. Ibid., 942.

71. Ibid., 944.

72. Ibid., 942.

73. Ibid.

74. The revival of political opposition in 1792 is most thoroughly discussed in Raymond Walters, Jr., "The Origins of the Jeffersonian Party in Pennsylvania," *Pennsylvania Magazine of History and Biography* 66 (1942), 440–58; Roland M. Baumann, "John Swanwick: Spokesman for 'Merchant-Republicanism' in Philadelphia, 1790–1798," ibid. 97 (1973), 148–56; and Harry M. Tinkcom, *The Republicans and Federalists in Pennsylvania, 1790–1801* (Harrisburg, Pa., 1950), 55–68.

75. On Hutchinson, see Whitfield J. Bell, Jr., "James Hutchinson (1752–1793): A Physician in Politics," in idem, *The Colonial Physician and Other Essays* (New York, 1975), 99–117.

76. On Hutchinson's "Old Whig" essays, see Chapter 3, above.

77. Walters, "Origins of the Jeffersonian Party," 443.

78. Edward C. Carter II, "A 'Wild Irishman' Under Every Federalist's Bed: Naturalization in Philadelphia, 1789–1806," *Pennsylvania Magazine of History and Biography* 94 (1970), 343; Hans-Jürgen Grabbe, "European Immigration to the United States in the Early National Period, 1783–1820," *Proceedings of the American Philosophical Society* 133 (1989), Table 1, p. 192; Maurice J. Bric, "Ireland, Irishmen, and the Broadening of the Late Eighteenth-Century Philadelphia Polity" (Ph.D. diss., Johns Hopkins University, 1990), Appendix II.2, 672.

79. Maldwyn A. Jones, "Ulster Emigration, 1783–1815," in E. R. R. Green, ed., *Essays in Scotch-Irish History* (London, 1969), 52–53; Bric, "Ireland," 299–307. On Irish immigration during the colonial era, see Sharon V. Salinger, *"To Serve Well and Faithfully": Labor and Indentured Servants in Pennsylvania, 1682–1800* (New York, 1987), 52–55; and Marianne S. Wokeck, "German and Irish Immigration to Colonial Philadelphia," *Proceedings of the American Philosophical Society* 133 (1989), 128–43.

80. The imigration totals are derived from Carter, "Wild Irishman," 341–43. The quote is from the *Belfast News Letter*, June 11, 1784, quoted in Jones, "Ulster Emigration," 51.

81. On the occupational composition of the immigrants, see Jones, "Uls-

ter Emigration," 50–55. The best discussion of Philadelphia's Irish weavers is Cynthia J. Shelton, *The Mills of Manayunk: Industrialization and Social Conflict in the Philadelphia Region, 1787–1837* (Baltimore, 1986). The quote is from J. Hamilton to Mr. Secretary Hamilton, Aug. 5, 1783, quoted in Jones, "Ulster Emigration," 50–51.

82. Carter, "Wild Irishman," 341–42.

83. R. G. Hill to Thomas Pelham, April 28, 1796, quoted in Jones, "Ulster Emigration," 55.

84. Thomas Ledlie Birch, *A Letter from an Irish Immigrant to His Friend in the United States* (Philadelphia, 1798). On the immigration of Irish radical leaders, see Jones, "Ulster Emigration," 55–56, and Richard Twomey, "Jacobins and Jeffersonians: Anglo-American Radicalism in the United States, 1790–1820" (Ph.D. diss., Northern Illinois University, 1974).

85. On Binns, see his autobiography, *Recollections of the Life of John Binns* (Philadelphia, 1854), and Twomey, "Jacobins and Jeffersonians," 24–27 and passim. On Carey, see Mathew Carey, *Autobiography* (Brooklyn, N.Y., 1942), originally published serially in 1833–34; and Kenneth W. Rowe, *Mathew Carey: A Study in American Economic Development* (Baltimore, 1933). On Duane, see Kim Tousley Phillips, "William Duane, Revolutionary Editor" (Ph.D. diss., University of California, Berkeley, 1968), and Chapters 5 and 6, below.

86. Tinkcom, *Republicans and Federalists*, 43.

87. On Sergeant, see Tinkcom, *Republicans and Federalists*, 54; Robert L. Brunhouse, *The Counter-revolution in Pennsylvania, 1776–1790* (Harrisburg, Pa., 1942), 79–80, 150–51.

88. McClenachan served as chairman of the Committee of Thirteen during the price-control controversy of 1779. See *Pennsylvania Packet* (Philadelphia), Aug. 12, 1779.

89. On McClenachan, see Tinkcom, *Republicans and Federalists*, 54; and Brunhouse, *Counter-revolution*, 214, 217–18.

90. Tinkcom, *Republicans and Federalists*, 55–62; Baumann, "John Swanwick," 148–56.

91. *American Daily Advertiser* (Philadelphia), July 27, 1792.

92. *General Advertiser* (Philadelphia), July 31, 1792.

93. James Hutchinson to Albert Gallatin, Aug. 19, 1792, Gallatin Papers, IV, New-York Historical Society, microfilm copy.

94. Ibid. Also see "Freedom of Election," *Independent Gazetteer*, Aug. 1, 1792.

95. Tinkcom, *Republicans and Federalists*, 64–65.

96. Roland M. Baumann, "The Democratic-Republicans of Philadelphia: The Origins, 1776–1797" (Ph.D. diss., Pennsylvania State University, 1970), chs. 6–8; idem, "John Swanwick," 148–49.

97. James Hutchinson to Albert Gallatin, Oct. 24, 1792, Gallatin Papers, IV, New-York Historical Society, microfilm copy.

98. Broadside, Oct. 9, 1792, quoted in Baumann, "John Swanwick," 149.

99. *To the Freemen of Pennsylvania* (Philadelphia, 1792).

100. Scharf and Westcott, *History of Philadelphia,* I, 468.

101. Bernard Fay, *The Two Franklins: Fathers of American Democracy* (Boston, 1933), 211–14; Charles Downer Hazen, *Contemporary American Opinion of the French Revolution* (Baltimore, 1897).

102. Scharf and Westcott, *History of Philadelphia,* I, 474.

103. On the formation of the Democratic Society, see Philip S. Foner, ed., *The Democratic-Republican Societies, 1790–1800: A Documentary Sourcebook of Constitutions, Declarations, Resolutions, and Toasts* (Westport, Conn., 1976), 3–51; and Eugene P. Link, *Democratic-Republican Societies, 1790–1800* (New York, 1942). The officers are listed in Scharf and Westcott, *History of Philadelphia,* I, 474.

104. Ibid., 468, 474.

105. On the role of religion in the formation of Philadelphia's early working class, see my "God and Workingmen: Popular Religion and the Formation of Philadelphia's Working Class, 1790–1830," in Ronald Hoffman and Peter J. Albert, eds., *Religion in a Revolutionary Age* (Charlottesville, Va., 1993). I discuss the intense competition between evangelicalism and traditional artisan values in "Alternative Communities: American Artisans and the Evangelical Appeal, 1790–1830," in Robert Asher, Howard Rock, and Paul Gilje, eds., *The American Artisan: New Perspectives* (forthcoming). For Bache's distribution of *The Age of Reason,* see James Tagg, *Benjamin Franklin Bache and the Philadelphia Aurora* (Philadelphia, 1991), 282.

106. The origin and early doctrines of American Universalism are discussed in Stephen A. Marini, *Radical Sects of Revolutionary New England* (Cambridge, Mass., 1982), 68–75; and Russell E. Miller, *The Larger Hope: The First Century of the Universalist Church in America, 1770–1870* (Boston, 1979).

107. Elhanan Winchester, *Thirteen Hymns, Suited to the Present Times,* 2d ed. (Baltimore, 1776), 15.

108. Universalist Church in the U.S.A., *Hymns Composed by Various Authors* (Walpole, N.H., 1808), 184–85. This, along with Winchester's *Thirteen Hymns,* formed the basic Universalist hymnody.

109. Some 39 percent of the church's subscribers appeared on the membership lists of the Democratic Society in 1793. This figure was arrived at by comparing the 1793 Subscription List of the First Universalist Church with the membership lists contained in the Minutes of the Democratic Society

of Pennsylvania. Both lists are at the Historical Society of Pennsylvania.

110. *Principles, Articles and Regulations Agreed upon by the Members of the Democratic Society in Philadelphia, May 30th, 1793* (Philadelphia, 1793).

111. Ibid.

112. Ibid.

113. Ibid., 65.

114. Ibid.

115. For the occupations of members of the Democratic Society, see Foner, *Democratic-Republican Societies*, Appendix I, 439–41.

116. The classic account of the epidemic is Mathew Carey, *A Short Account of the Malignant Fever Which Prevailed in Philadelphia in the Year 1793* (Philadelphia, 1793). For a modern account, see J. H. Powell, *Bring Out Your Dead: The Great Plague of Yellow Fever in Philadelphia in 1793* (Philadelphia, 1949).

117. Powell, *Bring Out Your Dead*, 96.

118. John Cox, *Rewards and Punishments, or, Satan's Kingdom Aristocratical* (Philadelphia, 1795).

119. Although some Federalists remained in the city, contemporaries remarked that Democratic-Republicans dominated relief activities. See Martin S. Pernick, "Politics, Parties, and Pestilence: Epidemic Yellow Fever and the Rise of the First Party System," *William and Mary Quarterly* 3d ser., 29 (1972), 577–80.

120. Of the known occupations of members of the Committee there were seven merchants, three carpenters, and one each of bookseller and printer, umbrella-maker, cabinetmaker, chair-maker, coachmaker, schoolmaster, cardmaker, cobbler and currier, and mechanic. See Powell, *Bring Out Your Dead*, 152–58.

121. The missing member of the correspondence committee was Alexander J. Dallas, whose position as Secretary of the Commonwealth required him to follow the state government into exile in Germantown. See Pernick, "Politics, Parties, and Pestilence," 578.

122. The meeting is described in Roland M. Baumann, "Philadelphia's Manufacturers and the Excise Taxes of 1794: The Forging of the Jeffersonian Coalition," *Pennsylvania Magazine of History and Biography* 106 (1982), 16–17.

123. Scharf and Westcott, *History of Philadelphia*, I, 478.

124. Ibid.

125. *Independent Gazetteer*, July 8, 1795.

126. Ibid.

127. Ibid.

128. Jacob Cox Parsons, ed., *Extracts from the Diary of Jacob Hiltzheimer of Philadelphia, 1765–1798* (Philadelphia, 1893), 215.

129. The events of July 4 are taken from *Independent Gazetteer*, July 8, 1795; *Aurora* (Philadelphia), July 9, 1795; and Douglas S. Freeman, *George Washington*, 7 vols. (New York, 1957), VII, 259.

130. Scharf and Westcott, *History of Philadelphia*, I, 485–86.

CHAPTER 5
AN APPRENTICESHIP TO CLASS

1. "Pittachus," *Aurora* (Philadelphia), Jan. 1, 1796. On Beckley's authorship of the "Pittachus" essay, see Edmund Berkeley and Dorothy Smith Berkeley, *John Beckley: Zealous Partisan in a Nation Divided* (Philadelphia, 1973), 120. On Beckley, see Philip S. Marsh, "John Beckley: Mystery Man of the Early Jeffersonians," *Pennsylvania Magazine of History and Biography* 72 (1948), 54–69.

2. *Aurora*, Jan. 9, 1796.

3. Ibid., July 6, 1796.

4. Ibid., Sept. 29, 1796.

5. Ibid., Oct. 6, 1796.

6. Ibid.

7. See Chapter 4, above.

8. *Aurora*, Oct. 11, 1796.

9. On the yellow fever epidemic of 1793, see Chapter 4, above.

10. *Aurora*, Oct. 11, 1796.

11. Ibid.

12. Ibid.

13. Ibid.

14. Ibid.

15. Ibid., Oct. 13, 15, 1796.

16. Ibid. On Swanwick and the merchant-Republicans, see Roland M. Baumann, "John Swanwick: Spokesman for 'Merchant-Republicanism' in Philadelphia, 1790–1798," *Pennsylvania Magazine of History and Biography* 97 (1973), 131–82.

17. *Aurora*, Oct. 19, 1796; John Beckley to Madison, Oct. 15, 1796, quoted in Baumann, "Swanwick," 174.

18. On Israel, see John K. Alexander, *Render Them Submissive: Responses to Poverty in Philadelphia, 1760–1800* (Amherst, Mass., 1980), 37–42.

19. *Aurora*, Oct. 8, 1797.

20. Ibid., Oct. 9, 1797.

21. Ibid.

22. Ibid., Oct. 10, 1797.

23. Ibid., Oct. 10, 13, 1797. Lists of those pledging relief money were published daily from this date until the fever subsided.

24. *Aurora*, Oct. 13, 1797.

25. Ibid., Oct. 14, 1797.

26. Ibid., Oct. 16, 1797.

27. Ibid.

28. Pennsylvania Senate Journal, 1797–98, quoted in Harry M. Tinkcom, *The Republicans and Federalists in Pennsylvania, 1790–1801* (Harrisburg, Pa., 1950), 178.

29. *Aurora*, Feb. 24, 1798.

30. Richard G. Miller, *Philadelphia—The Federalist City: A Study of Urban Politics, 1789–1801* (Port Washington, N.Y., 1976), Table 21, p. 150.

31. *Aurora*, Feb. 24, 1798.

32. *Porcupine's Gazette* (Philadelphia), Feb. 24, 26, 1798.

33. *Aurora*, Feb. 26, 1798.

34. Ibid., Feb. 27, 1798.

35. Ibid.

36. Ibid.

37. Ibid., Feb. 28, 1798.

38. Ibid.

39. Ibid., Feb. 27, 1798.

40. Ibid., Oct. 17, 1797.

41. Tinkcom, *Republicans and Federalists*, 238–39, 256–57.

42. On Duane's early life, see Kim Tousley Phillips, "William Duane, Revolutionary Editor" (Ph.D. diss., University of California, Berkeley, 1968), ch. 1.

43. On the London Corresponding Society and English Jacobinism in general, see E. P. Thompson, *The Making of the English Working Class* (New York, 1966), 17–21, 120–35, 153–74, 452–58; and Albert Goodwin, *The Friends of Liberty: The English Democratic Movement in the Age of the French Revolution* (Cambridge, Mass., 1979).

44. On Duane's activities in the London Corresponding Society, see Mary Thale, ed., *Selections from the Papers of the London Corresponding Society, 1792–1799* (Cambridge, Eng., 1983), 312–13, 322–25, 328–31.

45. Bache's paper appeared as the *General Advertiser* until late 1794 when he renamed it the *Aurora*. On Bache and the early history of his newspaper, see Jeffery A. Smith, *Franklin and Bache: Envisioning the*

Enlightened Republic (New York, 1990), chs. 5–7; and James Tagg, *Benjamin Franklin Bache and the Philadelphia Aurora* (Philadelphia, 1991).

46. On Eaton, see Thompson, *Making*, 97, 142.

47. *Aurora*, July 14, 1798. Such advertisements ran daily in the late 1790s.

48. Ibid., Dec. 13, 1798.

49. On postwar Irish immigration, see Edward C. Carter II, "A 'Wild Irishman' Under Every Federalist's Bed: Naturalization in Philadelphia, 1789–1806," *Pennsylvania Magazine of History and Biography* 94 (1970), 331–46; and Hans-Jürgen Grabbe, "European Immigration to the United States in the Early National Period, 1783–1820," *Proceedings of the American Philosophical Society* 133 (1989), Table 1, p. 192.

50. Phillips, "Duane," ch. 2. The history of the Alien and Sedition Acts is told in James Morton Smith, *Freedom's Fetters: The Alien and Sedition Laws and American Civil Liberties* (Ithaca, 1956).

51. Phillips, "Duane," ch. 4.

52. The story of the assault on Duane, which reveals perhaps better than anything else Federalist attitudes toward the city's laboring classes and its leaders, is told most fully in ibid., 70–75.

53. *Aurora*, April 21, 1800.

54. Ibid., May 13, 1800.

55. Ibid., Jan. 29, 1807. The italicized portion was an ironic reference to Samuel Johnson's *Dictionary of the English Language*, which had defined small producers in condescending terms as men of "mean, servile occupation employed in course work." See ibid., Feb. 3, 1807.

56. Ibid., Jan. 1, 1807.

57. On the contention between Philadelphia's Tory elite and the patriot laboring classes, see Chapter 2; on Arnold's treason, see John Richard Alden, *The American Revolution, 1775–1783* (New York, 1954), 209.

58. *Aurora*, Jan. 7, 1807.

59. Ibid., Jan. 29, 1807.

60. Ibid., Jan. 7, 1807.

61. Ibid., Jan. 15, 1807. On the contemporary debate over the place of commerce in the new republic, see Drew R. McCoy, *The Elusive Republic: Political Economy in Jeffersonian America* (Chapel Hill, 1980).

62. *Aurora*, Jan. 14, 1807.

63. Ibid., Jan. 15, 1807.

64. Ibid., Jan. 8, 1807.

65. Ibid., Jan. 29, 1807.

66. On the early English radical economists, see J. E. King, "Perish Commerce! Free Trade and Underconsumption in Early British Radical

Economics," *Australian Economic Papers* 20 (1981), 235–57; idem, "Utopian or Scientific? A Reconsideration of the Ricardian Socialists," *History of Political Economy* 15 (1983), 345–73; J. R. Dinwiddy, "Charles Hall, Early English Socialist," *International Review of Social History* 21 (1976), 256–76.

67. *Aurora*, Jan. 9, 1807.

68. Ibid., Jan. 12, 1807.

69. See Chapter 2, above.

70. *Aurora*, Feb. 2, 1807.

71. Ibid., Jan. 20, 1807.

72. Ibid., Feb. 5, 1807.

73. Ibid., Feb. 3, 5, 1807. It is probable that Duane took much of his analysis of English conditions from William Cobbett's *Political Register*. In the ninth of his essays he wrote that Cobbett was telling "the truth of *Duane*—now he is giving a just and true picture of the situation of the poor in England and of the middling tradesman." Ibid., Jan. 21, 1807. On Cobbett and his transformation from Federalist spokesman in Philadelphia to England's tribune of the poor, see Thompson, *Making*, 220–31, 452–71; and Raymond Williams, *Cobbett* (New York, 1983).

74. *Aurora*, Feb. 7, 1807. It is worth comparing Duane's ideas on the evils of unrestricted commerce with the congressional debate over maritime bankruptcy legislation at the turn of the century. Drew McCoy suggests that this debate centered around the theoretical question of the proper place of commerce in a republican society; Duane's arguments suggest something else, namely that the debate over commerce was, in fact, part of a larger struggle by small producers to maintain their social and economic position in the face of a rapidly rising mercantile class. See Chapter 4, above.

75. *Aurora*, Jan. 28, 1807. On the volatile issue of internal taxation, see Thomas P. Slaughter, "The Tax Man Cometh: Ideological Opposition to Internal Taxes, 1760–1790," *William and Mary Quarterly* 3d ser., 41 (1984), 566–91.

76. *Aurora*, Feb. 7, 1807.

77. Ibid., Feb. 5, 1807.

78. While a correspondent to John Binns's *Democratic Press* acknowledged the strike, he merely denounced the use of English common law in American courts. He had nothing to say concerning the journeymen or their employers. See "Antonius," *Democratic Press* (Philadelphia), May 4, 1807. The history of Philadelphia's important expatriate radical community is told in Richard Twomey, "Jacobins and Jeffersonians: Anglo-American Radicalism in the United States, 1790–1820" (Ph.D. diss., Northern Illinois University, 1974). Twomey claims that none of the expatriates, including Duane, supported the cordwainers' strike. While this may be true of the others, the

evidence presented here shows Duane to have been a strong and unwavering defender of the cordwainers.

79. John R. Commons, "American Shoemakers—A Sketch of Industrial Evolution, 1648–1895," *Quarterly Journal of Economics* (1909), incorporated in the Introduction to Volumes III and IV of John R. Commons and Helen L. Sumner, eds., A *Documentary History of American Industrial Society*, 10 vols. (Cleveland, Ohio, 1910), III, 19–58. On the cordwainers' trial, see Twomey, "Jacobins and Jeffersonians," ch. 6; and Ian M. G. Quimby, "The Cordwainers Protest: A Crisis in Labor Relations," *Winterthur Portfolio* III (1967), 83–101. The transcript of the trial is reprinted in Commons, *Documentary History*, III, 59–385.

80. *Aurora*, Nov. 27, 1805.

81. Ibid. Duane envisioned the prohibition of slave importations, due to take effect on Jan. 1, 1808, as leading directly to the emancipation of American slaves.

82. Ibid.

83. Ibid.

84. Ibid., Nov. 28, 1805.

85. Ibid.

86. Ibid., March 31, 1806.

87. Ibid.

88. Ibid.

89. Ibid., July 9, 1806.

90. Ibid.

91. Ibid., Feb. 7, 1807.

92. Ibid.

93. Ibid.

94. Ibid., March 27, 1807. On the English handloom weavers, see Thompson, *Making*, ch. 9; and Duncan Bythell, *The Handloom Weavers: A Study in the English Cotton Industry During the Industrial Revolution* (Cambridge, Eng., 1969).

95. *Aurora*, March 27, 31, 1807.

96. These and the following figures were computed from election slates printed every September and October in the *Aurora*. The occupations of candidates were determined from the annual city directories for Philadelphia.

97. *Aurora*, Oct. 12, 1812.

98. Ibid.

99. Ibid.

100. Susan E. Klepp, "Demography in Early Philadelphia, 1690–1860," *Proceedings of the American Philosophical Society* 133 (1989), Table 2, pp. 105–6.

101. Diane Lindstrom, *Economic Development in the Philadelphia Region, 1810–1850* (New York, 1978), 32–40.

102. Ibid., 33.

103. United States Census Office, 4th Census, 1820, *Records of the 1820 Census of Manufactures*, National Archives, returns for Philadelphia City and County, microfilm copy.

104. J. P. Brissot de Warville, *New Travels in the United States of America Performed in 1788* (Dublin, 1792), 325. On general economic trends in Philadelphia during the early industrial period, see Tom W. Smith, "The Dawn of the Urban-Industrial Age: The Social Structure of Philadelphia, 1790–1830" (Ph.D. diss., University of Chicago, 1980), ch. 2.

105. Tench Coxe, *An Essay on the Manufacturing Interest* (Philadelphia, 1804), 27.

106. On the Philadelphia Society, see David J. Jeremy, "The British Textile Technology Transmission to the United States: The Philadelphia Region Experience, 1770–1870," *Business History Review* 44 (1973), 29–32.

107. The story of Nicholson's industrial village, which failed, as did many early industrial ventures, of lack of capital and an inability to pay workers' wages promptly, is recounted in Cynthia J. Shelton, *The Mills of Manayunk: Industrialization and Social Conflict in the Philadelphia Region, 1787–1837* (Baltimore, 1986), ch. 1; and idem, "Labor and Capital in the Early Period of Manufacturing: The Failure of John Nicholson's Manufacturing Complex, 1793–1797," *Pennsylvania Magazine of History and Biography* 106 (1982), 341–64. Notices of the sale of mills and manufactories as well as notices of sheriff's sales of failed manufacturing establishments abound in the newspapers of this era.

108. McCoy, *Elusive Republic*, 104–19, 223–26, 243–48.

109. James Weston Livingood, *The Philadelphia-Baltimore Trade Rivalry, 1780–1860* (Harrisburg, Pa., 1947).

110. *Aurora*, Nov. 15, 1808.

111. *Aurora*, Nov. 21, 1808; *Democratic Press*, Nov. 22, 1808.

112. Tench Coxe, *A Statement of the Arts and Manufactures of the United States of America for the Year 1810* (Philadelphia, 1814).

113. Ibid.

114. Ibid.

115. Ibid.

116. Ibid.

117. Ibid.

118. Census Office, *1820 Census of Manufactures*.

119. Ibid.

120. On the depression of 1817–22, see Chapter 6, below.

121. These and the following figures were computed from the returns of

the 1820 manufacturing census. On the scale of producing units, see Smith, "Urban-Industrial Age," 40–45.

122. On the employment of women and children in America's early textile mills, see Shelton, *Mills of Manyunk*, ch. 3; Philip Scranton, *Proprietary Capitalism: The Textile Manufacture at Philadelphia, 1800–1885* (New York, 1983); Thomas Dublin, *Women at Work: The Transformation of Work and Community in Lowell, Massachusetts, 1826–1860* (New York, 1979); and Jonathan Prude, *The Coming of Industrial Order: Town and Factory Life in Rural Massachusetts, 1810–1860* (New York, 1983), 85–87, 211–22. On Philadelphia's early textile industry, see Smith, "Urban-Industrial Age," 32–40; and Adrienne Dora Hood, "Organization and Extent of Textile Manufacture in Eighteenth-Century, Rural Pennsylvania: A Case Study of Chester County" (Ph.D. diss., University of California, San Diego, 1988).

123. These and the following figures were computed from the returns of the 1820 manufacturing census. Cf., Smith, "Urban-Industrial Age," 45–67.

124. On the decline of household manufacturing in early nineteenth-century America, see Rolla M. Tryon, *Household Manufactures in the United States, 1640–1860* (Chicago, 1917).

125. On the city's machine-makers, see Coxe, *Statement of the Arts and Manufactures*, 48.

126. Census Office, *1820 Census of Manufactures*, return of Robert Wellford.

127. The primary source for information about the schism is Ezekiel Cooper, the Methodist book agent and a principle figure in the dispute. See his diary, as reprinted in George A. Phoebus, ed., *Beams of Light on Early American Methodism* (New York, 1887), 287–91. For a brief analysis of the schism from the perspective of Methodist Church history, see Doris Elisabett Andrews, "Popular Religion and the Revolution in the Middle Atlantic Ports: The Rise of the Methodists, 1770–1800" (Ph.D. diss., University of Pennsylvania, 1986), 301–15.

128. On the latent religiosity of the craft community and the alternatives offered by urban evangelicalism, see my "Alternative Communities: American Artisans and the Evangelical Appeal, 1790–1830," in Robert Asher, Howard Rock, and Paul Gilje, eds., *The American Artisan: New Perspectives* (forthcoming).

129. Andrews, "Popular Religion," 309–12.

130. [John Fanning Watson], *Methodist Error; Or Friendly Christian Advice, to Those Methodists Who Indulge in Extravagant Emotions and Bodily Exercises* (Trenton, N.J., 1819), 7–8.

131. Ibid., 8.

132. Ibid., 15, 8.

133. Philadelphia Hospitable Society, *The Nature and Design of the Hospitable Society* (Philadelphia, 1803), 3.

134. Ibid., 11.

135. Of those who could be located in city directories, cordwainers predominated. Only independent craftsmen and employing masters were listed in the directories.

136. Ibid., 5, 13.

137. Ibid., 11.

138. Such was the case among the more than 500 handloom weavers who found themselves rescued from poverty by the actions of the Domestic Society, which created employment for them in 1805. See James Mease, *The Picture of Philadelphia* (Philadelphia, 1811, rpt., 1970).

139. Hospitable Society, *Nature and Design*, 11.

140. This counterpoise between the rectitude and respectability of the upper ranks of the laboring classes and the noisy disrespectfulness of the lower ranks prefigures Bruce Laurie's description of revivalist and traditionalist cultures among Philadelphia working people in the second quarter of the nineteenth century. See Bruce Laurie, *Working People of Philadelphia, 1800–1850* (Philadelphia, 1980), chs. 2 and 3. On the rise of the "respectable" master artisan see Gary John Kornblith, "From Artisans to Businessmen: Master Mechanics in New England, 1789–1850" (Ph.D. diss., Princeton University, 1983).

141. This testimony is reprinted in Commons, *Documentary History*, III, 71–88.

142. Ibid., 72–73.

143. On early English friendly and trade societies, see John Rule, *The Experience of Labour in Eighteenth-Century Industry* (London, 1981), ch. 6; Thompson, *Making*, 418–23, 504–5, 513–15; and Clive Behagg, "Custom, Class and Change: The Trade Societies of Birmingham," *Social History* 4 (1979), 455–80.

144. Commons, *Documentary History*, III, 73.

145. See Chapter 4, above.

146. Commons, *Documentary History*, III, 74.

147. Ibid.

148. Ibid., 74–75.

149. Ibid., 77.

150. Ibid., 99, 101.

151. Ibid., 99–100, 78.

152. Ibid., 78–79.

153. Pullis was not the only craft leader to make this journey from friendly society to trade union. The executive committee of the journeymen

printers was divided over just this issue at the time of the cordwainers' strike. See Ethelbert Stewart, "A Documentary History of the Early Organization of Printers," in U.S. Bureau of Labor, *Bulletin 61* (Washington, D.C., 1905), 866.

CHAPTER 6
CONFRONTING INDUSTRIALISM

1. *Aurora* (Philadelphia), Sept. 21, 1810.
2. Ibid.
3. Ibid.
4. On ethnic divisions, especially between the established Scots-Irish and the Irish newcomers to the city, see Kim Tousley Phillips, "William Duane, Revolutionary Editor" (Ph.D. diss., University of California, Berkeley, 1968), 283–85; and Edward C. Carter II, "A 'Wild Irishman' Under Every Federalist's Bed: Naturalization in Philadelphia, 1789–1806," *Pennsylvania Magazine of History and Biography* 94 (1970), 331–46. On Dallas, see Raymond Walters, Jr., *Alexander James Dallas, Lawyer-Politician-Financier, 1759–1817* (Philadelphia, 1943).
5. *Aurora*, Sept. 21, 1810.
6. Ibid.
7. Ibid.
8. On the 1810 election, see Sanford W. Higginbotham, *The Keystone in the Democratic Arch: Pennsylvania Politics 1800–1816* (Harrisburg, Pa., 1952), 213–18, 245–46; and Phillips, "Duane," 308–14.
9. Anna Jacobson Schwartz, "The Beginning of Competitive Banking in Philadelphia, 1782–1809," *Journal of Political Economy* 55 (1947), 430–31.
10. On the merchant-Republicans, see Roland M. Baumann, "John Swanwick: Spokesman for 'Merchant-Republicanism' in Philadelphia, 1790–1798," *Pennsylvania Magazine of History and Biography* 97 (1973), 131–82; and idem, "Philadelphia's Manufacturers and the Excise Taxes of 1794: The Forging of the Jeffersonian Coalition," ibid. 106 (1982), 3–39.
11. On the Bank of the Northern Liberties in the election of 1810, see Kim T. Phillips, "Democrats of the Old School in the Era of Good Feelings," *Pennsylvania Magazine of History and Biography* 95 (1971), 366–67.
12. Mathew Carey, *Miscellaneous Essays* (Philadelphia, 1830), 255.
13. Ibid., 257.
14. Ibid., 259.
15. Louis Hartz, *Economic Policy and Democratic Thought: Pennsylvania, 1776–1860* (Cambridge, Mass., 1948), 69–81.

16. Ibid., 76.

17. Ibid., 74–75.

18. Ibid., 80.

19. Phillips, "Duane," 315–29, 475–78.

20. Stephen Simpson, *The Workingmen's Manual: A New Theory of Political Economy on the Principle of Production the Source of Wealth* (Philadelphia, 1831), 141–42.

21. Ibid., 142.

22. "National Finance," *Aurora*, Dec. 18, 1815, to April 10, 1816.

23. Phillips, "Duane," 475–78.

24. *Aurora*, Oct. 6, 1810.

25. *Aurora*, Oct. 15, 1810.

26. The results of the election are in Higginbotham, *Keystone*, 218.

27. *Aurora*, Oct. 12, 1810.

28. For a compelling account of the operations of garret masters in New York City, see Christine Stansell, "The Origins of the Sweatshop: Women and Early Industrialization in New York City," in Michael H. Frisch and Daniel J. Walkowitz, eds., *Working-Class America: Essays on Labor, Community, and American Society* (Urbana, Ill., 1983), 78–103. On garret shops in England, see E. P. Thompson, *The Making of the English Working Class* (New York, 1963), 253–68.

29. Alan Dawley, *Class and Community: The Industrial Revolution in Lynn* (Cambridge, Mass., 1976), 70.

30. Dorothy Thompson, *The Chartists: Popular Politics in the Industrial Revolution* (New York, 1984).

31. The best examples of Old School and New School ideas about industrialization are Stephen Simpson, *Workingmen's Manual*, for the Old School, and Mathew Carey, *Addresses of the Philadelphia Society for the Promotion of National Industry*, 5th ed. (Philadelphia, 1820), for the New School. Also see Carey's editorials in the *Political Economist* (Philadelphia) throughout 1824.

32. Mathew Carey, *An Appeal to Common Sense and Common Justice* (Philadelphia, 1822), 43.

33. For examples of the New School attitudes toward workers, see Carey, *Appeal to Common Sense*, as well as his *Addresses*.

34. Simpson, *Workingmen's Manual*, 172–73.

35. Ibid., 132.

36. The continued relevance of small-producer traditions in Philadelphia's early nineteenth-century political discourse argues against Steven Watts's claim that an ideological "shift toward liberal capitalism [w]as the crucial fact of early national life." Liberalism was indeed an emergent tradition between 1790 and 1820, but one that found its greatest support *outside*

the craft community. Emergent liberalism had little to say to the journeyman or ordinary workingman for whom the restoration of traditional small-producer values remained a real possibility and formed the driving force behind the nation's early working-class movements. See Steven Watts, *The Republic Reborn: War and the Making of Liberal America*, 1790–1820 (Baltimore, 1987); the quote is on page xvii.

37. *Aurora*, Oct. 19, 1812.

38. Higginbotham, *Keystone*, 264–65.

39. Ibid., 281; *Aurora*, March 23, 25; April 7, 1813; J. Thomas Scharf and Thompson Westcott, A *History of Philadelphia*, 1609–1884, 3 vols. (Philadelphia, 1884), I, 563.

40. For a view of the demand for labor and the high wages received, see *Minutes of the Committee of Defense of Philadelphia*, 1814–1815, in *Memoirs of the Historical Society of Pennsylvania* 8 (1867), 123.

41. Higginbotham, *Keystone*, 289; Scharf and Westcott, *History of Philadelphia*, I, 510–36. Also see the manuscript reports of the 1820 manufacturing census, United States Census Office, 4th Census, 1820, *Records of the 1820 Census of Manufactures*, National Archives, returns for Philadelphia City and County, microfilm copy.

42. *United States Gazette* (Philadelphia), Oct. 11, 1813, quoted in Higginbotham, *Keystone*, 289.

43. *Aurora*, Dec. 23, 25, 1813; Jan. 7, 1814.

44. The conduct of the war is discussed in Donald R. Hickey, *The War of 1812: A Forgotten Conflict* (Urbana, Ill., 1989); J. C. A. Stagg, *Mr. Madison's War: Politics, Diplomacy, and Warfare in the Early American Republic*, 1783–1830 (Princeton, 1983); Higginbotham, *Keystone*, chs. 10 and 11; and Victor A. Sapio, *Pennsylvania and the War of 1812* (Lexington, Ky., 1970).

45. On everyday life in the militia, see Thomas Franklin Pleasants, "Extracts from the Diary of Thomas Franklin Pleasants, 1814," *Pennsylvania Magazine of History and Biography* 39 (1915), 322–36, 410–24, for an officer's perspective; Florence Howard and Mary Howard, eds., "The Letters of John Patterson, 1812–1813," *Western Pennsylvania Historical Magazine* 23 (1940), 99–109, for a poignant account from a common soldier; and Committee of Defense, *Minutes*, 183.

46. See Howard and Howard, "Letters of John Patterson," 99–109; and the petition from the residents of the Northern Liberties to the Committee of Defense, Sept. 28, 1814 in Committee of Defense, *Minutes*, 188.

47. Committee of Defense, *Minutes*, 276.

48. Ibid., 297.

49. Ibid., 187–88.

50. *Aurora*, [n.d.]. Quoted in ibid., 15.

51. *Democratic Press* (Philadelphia), Jan. 13, 1813.

52. Ibid.

53. On Philadelphia's flourishing free black community, see Gary B. Nash, *Forging Freedom: The Formation of Philadelphia's Black Community, 1720–1840* (Cambridge, Mass., 1988).

54. 1820 Census of Manufactures, report #563.

55. Ibid., report #571.

56. Although the 1820 industrial census did not enquire as to unemployment, frequent comments by the respondents suggest that Philadelphia County manufacturers commonly laid off 50 to 75 percent of their workforce between 1817 and 1820. See ibid. One correspondent to the *Aurora* claimed the same degree of unemployment as early as 1816. See "The Farmers of Farm Hill," Jan. 6, 1816.

57. 1820 Census of Manufactures, report #563.

58. *Aurora*, Jan. 18, 1818.

59. Marion V. Brewington, "Maritime Philadelphia, 1609–1837," *Pennsylvania Magazine of History and Biography* 63 (1939), 111.

60. Unidentified source, no date, quoted in ibid.

61. *Aurora*, Jan. 6, 1816.

62. Ibid.

63. On the depressive effects of the end of the Napoleonic Wars, see Iorwerth Prothero, *Artisans and Politics in Early Nineteenth-Century London: John Gast and His Times* (Baton Rouge, La., 1979), 62–70; and Eric J. Hobsbawm, *Industry and Empire* (Harmondsworth, 1968), chs. 3–6.

64. Murray N. Rothbard, *The Panic of 1819: Reactions and Policies* (New York, 1962), ch. 1.

65. Phillips, "Duane," ch. 8, and 410–11. The best statement of Duane's postwar position on banking is contained in his essays entitled "National Finance," which appeared in the *Aurora* between Dec. 18, 1815, and April 10, 1816.

66. On the causes and effects of the Panic, see Rothbard, *Panic*, ch. 1.

67. *Democratic Press*, Jan. 1, 1819.

68. Ibid., Aug. 21, 1819.

69. Pennsylvania, General Assembly, House of Representatives, *Report*, Dec. 10, 1819.

70. Ibid., 3.

71. Ibid., 4.

72. Ibid.

73. Ibid., 4–5.

74. On the creation of the Society, see O. A. Pendleton, "Poor Relief in Philadelphia, 1790–1840," *Pennsylvania Magazine of History and Biogra-*

phy 70 (1946), 164–66; and Alan M. Zachary, "Social Disorder and the Philadelphia Elite Before Jackson," ibid., 99 (1975), 293–94.

75. Pendleton, "Poor Relief in Philadelphia," 165.

76. Pennsylvania Society for the Promotion of Public Economy, *Report of the Library Committee* (Philadelphia, 1817), 12.

77. Ibid., 14–15.

78. Pennsylvania, General Assembly, Senate, *Journal of the Senate of the Commonwealth of Pennsylvania, Session of 1820–21.* (Harrisburg, Pa., 1821), 334.

79. Ibid.

80. Pennsylvania House of Representatives, *Report*, 4.

81. On Leib's attempt, see Phillips, "Duane," 454–57.

82. Ibid., 456.

83. Ibid., 456–57.

84. Ibid.

85. Philip Klein, *Pennsylvania Politics, 1817–1832: A Game Without Rules* (Philadelphia, 1940), 48–49; the estimates of party strength are from the *Lancaster Weekly Journal*, March 17, 1820, quoted in ibid., 110.

86. *Aurora*, March 17, 1819.

87. Ibid., March 23, 1819.

88. Duane to Thomas Jefferson, Oct. 19, 1824, quoted in Phillips, "Duane," 590.

89. On the notion of "service," see Chapter 2, above.

90. *Equality* appeared between May 15 and July 3, 1802. The quote is from the *Temple of Reason* (Philadelphia), April 22, 1801.

91. For the authorship of *Equality* and a brief biography of Reynolds, see Richard Twomey, "Jacobins and Jeffersonians: Anglo-American Radicalism in the United States, 1790–1820" (Ph.D. diss., Northern Illinois University, 1974), 214–33.

92. For an account of the Duane-Reynolds trial, see James Morton Smith, *Freedom's Fetters: The Alien and Sedition Laws and American Civil Liberties* (Ithaca, 1956), 279–82.

93. James Reynolds, *Equality: A Political Romance* (Philadelphia, 1802), xxii.

94. Ibid., 8.

95. Ibid., 8–9.

96. For Reynolds's participation in Old School politics, see Phillips, "Duane," 286. Reynolds's name appeared frequently in announcements of party affairs carried in the *Aurora* in this period.

97. Duane's essays are discussed in Chapter 5, above.

98. *Aurora*, Jan. 15, Feb. 7, 1807.

99. On Maclure see Arthur Bestor, *Backwoods Utopias: The Sectarian Origins and the Owenite Phases of Communitarian Socialism in America, 1663–1829*, 2d ed. (Philadelphia, 1970), 146–53; and Anthony F. C. Wallace, *Rockdale: The Growth of an American Village in the Early Industrial Revolution* (New York, 1978), 263–70.

100. William Maclure, *Opinions on Various Subjects Dedicated to the Industrious Producers* (Philadelphia, 1831), quoted in Bestor, *Backwoods Utopias*, 149.

101. Wallace, *Rockdale*, 266–68.

102. Quoted in ibid., 267.

103. Cornelius C. Blatchly, *Some Causes of Popular Poverty*, appended to Thomas Branagan, *The Pleasures of Contemplation* (Philadelphia, 1817), 192–220. On Blatchly, see David Harris, *Socialist Origins in the United States: American Forerunners of Marx, 1817–1832* (Assen, The Netherlands, 1966), 10–19.

104. Blatchly, *Popular Poverty*, 197.

105. On the history of this idea and its importance in the socialist movement, see A. Menger, *The Right to the Whole Produce of Labour* (London, 1899).

106. Blatchly, *Popular Poverty*, 198-99.

107. Ibid., 199.

108. Ibid.

109. The quote is from ibid., 210.

110. Ibid., 204–5.

111. Ibid., 204.

112. Ibid., 207.

113. The rise of laboring-class religion in Philadelphia is discussed in my "God and Workingmen: Popular Religion and the Formation of Philadelphia's Working Class, 1790–1830," in Ronald Hoffman and Peter J. Albert, eds., *Religion in a Revolutionary Age* (Charlottesville, Va., 1993).

114. On Simpson, see Edward Pessen, "The Ideology of Stephen Simpson, Upper-Class Champion of the Early Philadelphia Workingmen's Movement," *Pennsylvania History* 22 (1955), 328–40; idem, *Most Uncommon Jacksonians: The Radical Leaders of the Early Labor Movement* (Albany, N.Y., 1967), 75–78; and Joseph Dorfman, *The Economic Mind in American Civilization, 1606–1865*, 2 vols. (New York, 1946), II, 645–48.

115. Phillips, "Duane," 463–65.

116. Simpson, *Workingmen's Manual*, 23.

117. Ibid., 29.

118. Ibid., 70.

119. Ibid.

120. Ibid., 48, 69.

121. Reginald C. McGrane, ed., *The Correspondence of Nicholas Biddle Dealing with National Affairs, 1807–1844* (Boston, 1919), 14–15.

122. Diane Lindstrom, *Economic Development in the Philadelphia Region, 1810–1850* (New York, 1978); idem, "The Industrial Revolution in America," in Sidney Pollard, ed., *Region und Industrialisierung: Studien zur Rolle der Region in der Wirtschaftsgeschichte der Letzten Zwei Jahrhunderte* (Gottingen, 1980), 69–88; Thomas C. Cochran, "Philadelphia: The American Industrial Center, 1750–1850," *Pennsylvania Magazine of History and Biography* 106 (1982), 323–40.

123. On the 1820 election, which was organized by Simpson, see Phillips, "Duane," 544–46; the election results are in the *Aurora*, Nov. 6, 1820.

124. For Duane's protest campaign, see Phillips, "Duane," 529.

125. The meeting is discussed in ibid., 544.

126. Ibid., 529.

127. On the "original Jacksonians," see ibid., 548–56, 576–88; idem, "Democrats of the Old School," 381–82.

128. John Lisle to George Bryan, Jan. 25, 1823, quoted in Phillips, "Duane," 450.

CHAPTER 7
Making the Republic of Labor

1. On Owen, see J. F. C. Harrison, *Quest for the New Moral World: Robert Owen and the Owenites in Britain and America* (New York, 1969); Arthur Bestor, *Backwoods Utopias: The Sectarian Origins and Owenite Phases of Communitarian Socialism in America, 1663–1829*, 2d ed. (Philadelphia, 1970), ch. 4; and E. P. Thompson, *The Making of the English Working Class* (New York, 1966), 779–806.

2. Robert Owen, *A New View of Society* (London, 1817, rpt., 1968).

3. *New Harmony Gazette*, Oct. 1, 1825, quoted in Harrison, *New Moral World*, 59.

4. Thomas Branagan, *The Pleasures of Contemplation* (Philadelphia, 1817), 221.

5. Bestor, *Backwoods Utopias*, 100.

6. On Owen's activities in Philadelphia and his influence in America, see Bestor, *Backwoods Utopias*, 96–97, 100, 107–8, 110–11, 128–29; W. G. H. Armytage, "Owen in America," in Sidney Pollard and John Salt, eds., *Robert Owen, Prophet of the Poor: Essays in Honor of the Two Hundredth Anniversary of His Birth* (London, 1970), 214–38; and Harrison, *New Moral World*.

7. Bestor, *Backwoods Utopias*, 152–59, 202–3; Anthony F. C. Wallace,

Rockdale: The Growth of an American Village in the Early Industrial Revolution (New York, 1978), 284–89. On the social composition of the commune, see Wallace, ibid., 286–87. The middle-class nature of the projected Valley Forge community is confirmed by the daughter of one of its principal organizers. See Jane D. Knight, *Events Touching Various Reforms* (Albany, N.Y., 1880), 17–19.

8. On Thompson and Gray, see J. E. King, "Utopian or Scientific? A Reconsideration of the Ricardian Socialists," *History of Political Economy* 15 (1983), 345–73; J. Kimball, *The Economic Doctrines of John Gray, 1799–1882* (Washington, D.C., 1948); and Maurice Dobb, *Theories of Value and Distribution Since Adam Smith* (Cambridge, Eng., 1973), 137–47.

9. King, "Utopian or Scientific?," 350.

10. Harrison, *New Moral World*, 64.

11. Bestor, *Backwoods Utopias*, 149.

12. John Gray, *A Lecture on Human Happiness* (Philadelphia, 1825), 5.

13. Ibid.

14. Ibid., 6.

15. Ibid., 15.

16. Ibid., 26–27.

17. Ibid., 35–36.

18. Ibid., 37, 39.

19. Ibid.

20. Ibid., 61.

21. Ibid., 63.

22. Ibid., 62.

23. Ibid., 59–71.

24. Ibid., 68.

25. On the popular beliefs that made up "traditional Christianity," see John Bossy, *Christianity in the West, 1400–1700* (New York, 1985). Compare Bossy's argument with that of Jon Butler, who views popular beliefs prior to the Second Great Awakening as fundamentally non-Christian in nature. See his *Awash in a Sea of Faith: Christianizing the American People* (Cambridge, Mass., 1990). My own preliminary findings suggest that American artisans were far more Christianized than Butler allows. See my "Alternative Communities: American Artisans and the Evangelical Appeal, 1790–1830," in Robert Asher, Howard Rock, and Paul Gilje, eds., *The American Artisan: New Perspectives* (forthcoming).

26. Frank Baker, *From Wesley to Asbury: Studies in Early American Methodism* (Durham, N.C., 1976), 32.

27. These figures are taken from my "God and Workingmen: Popular Religion and the Formation of Philadelphia's Working Class, 1790–1830,"

in Ronald Hoffman and Peter J. Albert, eds., *Religion in a Revolutionary Age* (Charlottesville, Va., 1993), Table 1.

28. The total membership of Philadelphia's Methodist churches in 1794 was only 365; in 1801, following one of the city's most successful revivals, the number was still a mere 669. These figures include both men and women as well as the city's black congregants and are taken from Doris Elisabett Andrews, "Popular Religion and the Revolution in the Middle Atlantic Ports: The Rise of the Methodists, 1770–1800" (Ph.D. diss., University of Pennsylvania, 1986), Table I, 171.

29. George Peck, *Early Methodism within the Bounds of the Old Genessee Conference from 1788 to 1828* (New York, 1860), 120.

30. Ibid., 73.

31. Ibid., 151, 167.

32. Robert Collyer, *Father Taylor* (Boston, 1906), 39.

33. On the triumph of evangelicalism in the 1830s and 1840s, see Bruce Laurie, *Working People of Philadelphia, 1800–1850* (Philadelphia, 1980).

34. Louis H. Arky, "The Mechanics' Union of Trade Associations and the Formation of the Philadelphia Workingmen's Movement," *Pennsylvania Magazine of History and Biography* 76 (1952), 15–19.

35. On the creation of new journeymen's societies in the 1820s, see William A. Sullivan, *The Industrial Worker in Pennsylvania, 1800–1840* (Harrisburg, Pa., 1955), Appendix B, 221; Arky, "Mechanics' Union," 159; and John R. Commons et al., *History of Labour in the United States*, 2 vols. (New York, 1936), I, 157–58; Thomas Porter, *Picture of Philadelphia* (Philadelphia, 1831), 267–76.

36. Sullivan, *Industrial Worker*, Appendix B, 221.

37. Ibid., 221-23.

38. For examples of this view, see Commons, *History of Labour*; Philip S. Foner, *History of the Labor Movement in the United States*, 8 vols. (New York, 1947–   ), I; and Edward Pessen, *Most Uncommon Jacksonians: The Radical Leaders of the Early Labor Movement* (Albany, N.Y., 1967).

39. On Heighton and the working-class movement in Philadelphia, see Louis H. Arky, "The Mechanics' Union of Trade Associations and the Formation of the Philadelphia Workingmen's Movement" (Ph.D. diss., University of Pennsylvania, 1952); idem, "Mechanics' Union"; Laurie, *Working People of Philadelphia*, 75–86; and Pessen, *Most Uncommon Jacksonians*, chs. 1 and 2.

40. Arky, "Mechanics' Union," 144.

41. Heighton had the highest regard for Gray's *Lecture*, writing at one point that "this little work . . . contains truths that could not fail to open the eyes of mankind; it ought to be read by every man." *Address to the*

Members of Trade Societies and to the Working Classes Generally (Philadelphia, 1827), 15.

42. Ibid., 3–4.

43. Ibid., 9–15.

44. Ibid., 15–19; Heighton's comment appears on 19.

45. Ibid., 4, 7.

46. Ibid., 31.

47. Ibid., 34–35.

48. Ibid., 35.

49. Ibid., 12.

50. Arky, "Mechanics' Union," 152–54.

51. In addition to the *Address to Trade Societies*, Heighton printed two more of the speeches that he delivered among Philadelphia's working classes: [William Heighton], *An Address Delivered Before the Mechanics and Working Classes Generally, of the City and County of Philadelphia* (Philadelphia, 1827); and [William Heighton], *The Principles of Aristocratic Legislation* (Philadelphia, 1828).

52. Heighton, *Address to Trade Societies*, 37.

53. Heighton, *Principles of Aristocratic Legislation*.

54. Heighton, *Address to Trade Societies*, 37.

55. See Chapter 5, above.

56. Heighton, *An Address Delivered*, 9–10.

57. On the laboring-class orientation of Philadelphia Universalism, see my "God and Workingmen."

58. These figures were reached by comparing the 1793 Subscription List and the 1814–25 Pew Book of Philadelphia's First Universalist Church with city directories for those years. All can be found at the Historical Society of Pennsylvania.

59. Hosea Ballou, Abner Kneeland, and Edward Turner, *Hymns Composed by Different Authors* (Walpole, N.H., 1808), 157. This was a standard collection of Universalist hymns used throughout the United States.

60. Ibid., 160.

61. Ibid., 7.

62. Heighton, *An Address Delivered*, 9.

63. Ibid.

64. Ibid.

65. Ibid.

66. Ibid.

67. Ibid.

68. Ibid.

69. Ibid.

70. Ibid.

71. *Aurora*, Oct. 25, 1828.

72. Ibid.

73. *Mechanics' Free Press* (Philadelphia), Oct. 25, 1828.

74. *Mechanics' Free Press*, April 26, May 3, 10, 1828. The producer and consumer cooperatives of the 1840s and 1850s have not received the attention they deserve. For a brief analysis, see Laurie, *Working People of Philadelphia*, 102–3, 193–94.

75. Conspicuously absent from the pages of the *Mechanics' Free Press* were discussions of religion. Faced with a potential division of the workingmen's movement into denominational factions, Heighton urged his readers to "let the subject of religion alone." *Mechanics' Free Press*, Dec. 17, 1829.

76. Heighton, *Principles of Aristocratic Legislation*, 15.

77. On the history of the Workingmen's party, see Arky, "Mechanics' Union," 161–74; and Pessen, *Uncommon Jacksonians*, 3–51.

78. Heighton, *Principles of Aristocratic Legislation*, 6.

79. Ibid., 7.

80. Kim Tousley Phillips, "William Duane, Revolutionary Editor" (Ph.D. diss., University of California, Berkeley, 1968), 610–15.

81. *Mechanics' Free Press*, Sept. 27, 1828; March 21, 1829.

82. Phillips, "Duane," 611–13.

83. John R. Commons and Helen Sumner, eds., *A Documentary History of American Industrial Society*, 10 vols. (Cleveland, Ohio, 1910), V, 76.

84. *Mechanics' Free Press*, Oct. 18, 1828.

85. On Jackson's popularity among Philadelphia workingmen, see Phillips, "Duane," ch. 15.

86. Ibid., 605–10.

87. Ibid., 607.

88. *Mechanics' Free Press*, Oct. 17, 1829. The Working Men's Association of Southwark was even more enthusiastic than Heighton. Shortly after the election, the Association declared that "the objects we have in view is the finish of the glorious work of the Revolution." Ibid., Oct. 31, 1829.

89. Leonard Bernstein, "The Working People of Philadelphia from Colonial Times to the General Strike of 1835," *Pennsylvania Magazine of History and Biography* 74 (1950), 332–33.

90. *Mechanics' Free Press*, Oct. 16, 1830. On the corruption of the Jackson machine, see Phillips, "Duane," 614–15.

91. Bernstein, "Working People of Philadelphia," 332–33.

92. William Heighton to George L. Stearns, Feb. 27, 1865, in [Anon.], *The Equality of All Men before the Law Claimed and Defended* (Boston, 1865), 42–43. The only William Heighton listed in the 1840 and 1850 federal manuscript censuses resided in the state of Indiana.

EPILOGUE

1. The quote is from John Ferral, speaking at the 1834 convention of the National Trades Union, in John R. Commons and Helen L. Sumner, eds., *A Documentary History of American Industrial Society*, 10 vols. (Cleveland, Ohio, 1910), VI, 216.

2. Ibid., 211.

3. Ibid.

4. Ibid., 215.

5. Ibid., 215–16.

6. Ibid., 212–13.

7. Ibid., 214.

8. For Heighton, see *Mechanics' Free Press* (Philadelphia), Oct. 17, Nov. 14, 1829; for English, see Commons, *Documentary History*, VI, 215.

Index

Academy of Natural Sciences, 212
Adcock, William, Constitutional Society leader, 54
Address of John Humble, 80
Advice to the Privileged Orders, 113
African-Americans, workers during Second Anglo-American War, 194–195
Age of Reason, distributed, 131
Agreement of the People, 11–13, 43
Alcorn, George, Democratic Cordwainers' leader, 164
Alien and Sedition acts, signed, 139
Almshouse admissions (1780s), 116
American Revolution: artisans joining the, 37; meaning of to artisans, 37
Anti-Federalists, 77, 79 (*see also* Constitutionalists); among laboring class, 83, 84; development of movement, xiv; Gallatin on view of toward merchants, 115; in ratification controversy, 82–83; small-producers' republic favored by, 113; views of on French Revolution, 130
Apprenticeship programs, changes in, 40
"Arminius" letter, 97–98
Arnold, Benedict, Philadelphia connection of, 155
Articles of Confederation, 77; critics' contrasting to Constitution, 81
Artisan socialism, xi
Artisan thought, tradition of, xii, xiii
Artisans: as bedrock of community, 5; benefits of organizing, 118; discontent, 92–93; in economic depression and revival, 117–118; organization of, 118

Associators militia, 28; equalized participation demanded by, 48; first combat of, 46; grievances, 47; membership, 45–46; morale of, 46
Aurora (newspaper), xiv; as radical journal, 152; contents, 152–153; edited by Bache, 152; edited by Duane, 152; on postwar depression, 196; on rise of industry, 166; Owen writings in, 212; reports on Fries Rebellion in, 153; silence of in 1810 election, 187

Bache, Benjamin Franklin: distributes Paine tract, 131; on Israel Israel, 146; on Israel Israel defeat, 148; on Israel Israel election, 147; radical books sold by, 152; support for United Irish, 152
Baltimore, as competing port, 166
Bank of North America: complaints against, 85; formation of, 84; interest rates of, 85; loss of charter, 86; monopoly control of, 84–85; rechartering, 87; stockholders in, 84
Bank of Northern Liberties: debate over as divider of labor, 184, 187; organized, 184
Bankruptcy law: anti-Federalists' opposition to, 115; debates over, 114
Banks: as partisan institutions, 183; boom and collapse of, 196–197; chartering, 196; collapse in Panic of 1819, 197; Duane views on, 186; Mathew Carey on, 184; organized by Quids, 184; politi-

Banks (*Cont.*)
cal debate over, 184; regulation of, 184
Barlow, Joel, support for Webster's program by, 113
Bastille Day celebration (1794), 137
Bayard, James A., on side of merchants, 115
Beckley, John, Democratic-Republican leader, 142
Bedford, John, 176–178; resistance of to Journeymen Cordwainers Society, 177
Bengal Journal, edited by William Duane, 151
Biddle, Charles, founder of Democratic Society of Pennsylvania, 130
Biddle, Nicholas, on city politics, 209
Bingham, Anne Willing, 109
Bingham, William: pamphlet by on West Indies trade, 76; wealth of, 109
Binns, John, 125; newspaper owned by, 194
Black Bear tavern, pro-French Revolution displays at, 130
Blackwell, John, appointed Penn. governor, 18
Blatchly, Cornelius, 205; advocates Christian commonwealth, 207; Gray compared to, 215; socialist theories of, 205–207
Bond, Phineas, on impoverished laboring classes, 116
Book of Prices, 119; as assertion of artisan rights, 120
Boycotts, by cordwainers against imported shoes, 98
Boyd, Alexander, refusal to join reconstituted Committee of Privates, 58
Branagan, Thomas, on Owen's ideas, 212
Bread, scarcity (1779), 53
Breweries, 167–168
Brewster, Walter, on labor as property, 5

Bright, Jacob, council candidate, 142
"Brother Mechanic," address by, 96–97
Brown, Arthur Brockden, account of 1793 yellow fever epidemic, 134
Bryan, George: on constitutional convention, 77; opposition to nationalizing trade, 77
Bull, John, refusal to join reconstituted Committee of Privates, 58
Bull, William, cloth-maker heretic, 8; on religion, 8, 9

Cabinet and chair-makers: form own organization, 123; issue Book of Prices, 119
Cannon, James: Committee of Privates leader, 34; moves from Philadelphia, 58
Capitalism (*see also* Socialism; Small-producer tradition): rise of, generally, xii
Capitalist manufacturing, rise of, xiii; *see also* Manufacturing
Carey, Mathew, 125; as merchant-Republican leader, 183; guides Owen, 213; on bank wars, 184; praises domestic manufacturers, 190; statistical account of yellow fever epidemic by, 134, 135
Carpenter, Samuel, hard money legislator, 21
Caucus system, critique of by socialists, 224
Central trade union, xv
Chamber of Commerce, call for by Pelatiah Webster, 111
Charter, corporation (*see also* Corporation): colonial, 87; features of, 89; restoration, 88–89
Cigar manufacture, 168
Clarkson, Matthew: as alderman, 89; as establishment leader, 88; as mayor during yellow fever epidemic, 135, 136; on merchant-craftsman committee (1793), 136
Cloth trade, 7, 8
Coates, Samuel, notebooks of, 90

Cobbett, William, on Israel Israel defeat, 148
Collier, Richard, report by on refusal to repair imported shoes, 98
Commerce: debate on place of in American society, 113; revival of (1790s), 117–118; with West Indies advocated, 76
Commercial regulation (*see also* Regulation): following ratification of U.S. Constitution, 84; support for federal control of, 76, 77
Committee, The, *see* The Committee.
Committee of Defense, relief for soldiers' families by, 194
Committee of Inspection and Observation: ouster of conservatives from, 33; town meeting called by, 41
Committee of Privates, 30–31; educative function of, 34; leaders, 34; membership, 33; Memorial of (1776), 34; reconstituted, 58; suffrage goals of reached, 42
Committee of Thirteen, 64, 65; on community, 66–67; review of history of, 65–66
Committee of Trade: calls for price controls, 57; citizens' plan, 57; election of, 55; dissolution, 57; importance of, 57; warnings issued by, 56
Community: Committee of Thirteen view of, 66–67; craftsmen's commitment to, 7; idea of, 66–67
Competency: as a craftsman's goal, 6; definition of, 7
Constitution of 1776 (Revolutionary Constitution): celebration of, 104; conservative opposition to, 53; Constitutionalist support for, 71; Declaration of Rights, 44; provisions, 42–44; Republican attacks against, 73–74; revised, 75; term limits in, 43; tone of, 44
Constitution, U.S.: laboring-class support for, 83, 84; ratification, 84, 107

Constitution of 1790, provisions, 75
Constitutional convention: call for (1776), 42; state (1789), 74–75, 77
Constitutional Society: call for town meeting, 54; city divisions shown in formation of, 58; formation of, 53; leaders, 54
Constitutionalist Party, dissolution, 123
Constitutionalists: attention of to workingmen's issues, 98; attitudes of laborers toward, 96; domination by rural interests of, 83; goals, 71; in postwar years, 70; isolation of to 1792, 115–116; leaders, 69; opposition to Bank of North America, 86; opposition to Republicans, 70–71; response to Republican attacks on Constitution, 73; restrictions by on rechartered bank, 87; view of government, 71
Consumer associations, formation of, 193
Consumer goods, manufacture of, 167
Continental Congress, memorial of Committee of Privates to, 34
Cordwainers, journeymen (*see also* Journeymen Cordwainers' Society): price of regulation by, 120; strike, 150, 160–162, 176–178
Cordwainers' strike (1805), 150, 160; Duane on, 160–162; Harrison testimony about, 176–178; John R. Commons on, 160; trial, 160
Corporation, Philadelphia City: anti-labor practices of, 144; attacks by Democratic-Republicans on, 143; ban on affordable housing, 108; collapse of in Revolution, 88; fire protection advocated by, 108; fiscal practices questioned, 143; food-purity rules issued by, 108; in colonial period, 87; leadership, 87–88; reassertion of elite authority through, 107–108; regulatory ordinances of, 143–144;

Corporation, Philadelphia City
(*Cont.*)
  restoration of, 88–89; review of
  by drayman, 108
Council of Censors: election of 1782,
  72; in revolutionary constitution,
  43
Cox, John, poem by on yellow fever
  epidemic, 135
Coxe, Tench: as merchant-Republican leader, 183; industrial
  census by, 167–168; on Philadelphia economy, 165
Craft production, decline of, xiii,
  xiv
Craftsmen (*see also* Small-producer
  tradition): complaints of in Revolution, 52–53; discontent, 51,
  52; monopoly of skills, 169
Credit: controlled by Bank of North
  America, 84–85; from General
  Loan Office, 85–86; requirement
  for in international trade, 84
Cromwell, Oliver, and Levellers, 13
Crosby, James, on postwar depression, 195
Cross-Keys Tavern, as political meeting place, 145
Cuthbert, Anthony: council candidacy, 142; on non-importation
  committee, 100

Dallas, Alexander James: and New
  School Democrats, 182; appointed Secretary of Commonwealth, 125; as member of
  Democratic Society of Pennsylvania, 131; edits *Pennsylvania
  Evening Herald*, 125; Federalist
  opposition leader, 125; on naming party, 128; on town meeting
  (1792), 127
Declaration of Rights, in revolutionary constitution, 44
Deism, contrasted to Universalism,
  131
Dell, William, thought paralleled by
  William Keith, 22
Democratic Cordwainers: election

day march by, 164; organized,
  163; support of for Democratic-
  Republicans, 164
Democratic Press, 194; calls for wage
  regulation, 197
Democratic Society of Pennsylvania:
  craftsmen in, 134; Democratic-
  Republican leaders as members
  of, 131; following epidemic, 137;
  formed, 130; in epidemic, 136;
  leaders, 130–131; membership,
  133–134; principles, 132–133;
  radical wing of, 131; Thomas
  Paine and, 132
Democratic-Republican Party: alliance with workingmen by, xiv;
  attacks against Corporation by,
  143; attention of toward laboring
  class, 142, 143; candidates
  (1796), 142–143; coalition of
  with labor, 150; coalitions in,
  183; council tickets, 163; in
  election of 1796, 144–145; Israel
  Israel election, 147; labor wing
  of, xv; leaders of in Democratic
  Society of Pennsylvania, 131;
  merchant-Republican wing, 183;
  split of into Old and New
  School Democrats, 182; survival
  of coalition, 151; turning point
  for, 149; victories (1798–1800),
  151; workingmen support for
  (1792), 128
Depression (1722), 20; views of
  causes of, 20–21
Depression (1760), 29, 38
Depression (1785), 76, 116
Depression (1820), 168, 195
De Warville, Brissot, on Philadelphia
  economy, 165
Dickens, A. G., on William Bull, 8
Distilleries, 167–168
Drinker, John, seized, 59
Duane, William, xiv, 125, 221; adju-
  tant general of militia (1813),
  192; arguments of recast by
  Blatchly, 206; as friend of Dr.
  James Reynolds, 202; as Old
  School leader, 185; as social

commentator and critic, 200; as
Workingmen's Party member,
230; attacks merchants, 157;
banking views, 186; career, 151–
153; contribution of to socialism,
203–204; cordwainers' tributes
to, 163–164; doctrines espoused
by, 153–154, 155; editor of *Aurora*, 152; editor of *Bengal Journal*, 151; effect of on Democratic-Republican council
tickets, 163–164; influence of on
Workingmen's Party, 230; inves-
tigates economic conditions,
197–198, 199; Irish sympathies
of, 153; leader of remnants of
Old School, 201; on banking ex-
pansions, 196–197; on caucus
system, 224; on class conflict,
162–163; on commerce and
trade, 156; on cordwainers'
strike, 160–162; on 1810 election results, 187, 188; on relief
for soldiers' families, 194; on
small-producers' tradition, 153–
159; on value of labor, 156; program of, 159; propaganda of as
unifier, 181; prosecuted for sedition, 202; publishes Owen writings, 212; retirement of, 233;
support for Neff's school, 205
Duane, William John: as Andrew
Jackson's treasury secretary,
233; in 1820 election, 210;
leader of remnants of Old
School, 201
Duane-Leib machine, 182
Du Ponceau, Peter, founder of Democratic Society of Pa., 130

Eaton, Daniel Isacc, Bache as agent
for, 152
Economic change, in Philadelphia,
164–165
Economic recovery (1790), 117
Election of 1783, workingmen's
course in, 97
Election of 1784, imports as issue in,
97

Election of 1789, first under new
charter, 89
Election of 1792: attempts to reunite
middle-class Democrats and arti-
sans in, 126; results, 128
Election of 1796, 144–145
Election of 1810, xv, 181–187; im-
portance of, 188–190; results,
187; splits leading up to, 182–
187
Election of 1812, 192
Elites: haughtiness of Philadelphia,
109; in early federal period, 108,
109
Emerson, James, 217
English Civil War: and small-produc-
ers' tradition, 10–14; legacy, 14
English Combination Acts, 160
English social criticism, in William
Duane writings, 154
Equality: A Political Romance (so-
cialist tract), 202
Evangelical movement, 218–220
Evangelists, workingmen and, 211
Evans, Edward, 217
Eyre, Emanuel, Democratic-
Republican candidate, 142

Federal Procession of 1788, 82
Federalist Papers: attack on small-
producer tradition in No. 10,
107; on liberalism, 105
Federalists: appeals to labor to support federal constitution, 80–81;
as dominant culture of Philadelphia, 109–110; attack on, 79,
80; call for statewide nominating
convention (1792), 126; criticism
of for federal constitution terms,
79; desertion during yellow fever
epidemic, 135–136; election irregularities of, 149; in debate
over small-producers' role, 115;
intimidation of voters by, 148;
isolated from events, 130; laboring-class support for (1810),
181; nominating convention
(1792), 128; opposition to, 123;
program, 81; support for ratifica-

Federalists (*Cont.*)
tion of U.S. Constitution, 82;
workingmen support for, 82
First Universalist Church (*see also*
Universalism): and artisan
values, 131; as a political gathering place, 132; labor members
of, 225
Fisher, Miers: establishment leader,
88; on council, 89
Fox, George, Quaker founder, 15
Food prices (1770s), 38
Fort Wilson riot, 51, 59–60; aftermath, 60; broadside posted after,
92; militia defeat in, 69; recalled, 99, 123
Foreclosures, 199; *see also Depression
entries*
Frame of Government (William
Penn's), 16; terms, 16; rejection
of, 18
Franklin, Benjamin: and Stamp Act,
31; cited by Duane on small-
producer tradition, 155–156;
Junto meetings of, 27; negotiating royal charter, 31, 32; support
of for paper money, 25
Franklin Institute, Owen lectures to,
213
Franklin Typographical Society:
formed, 119; functions, 119;
strike by, 119, 120
Free trade, Pelatiah Webster views
on, 112
French Revolution: demonstrations in
favor of, 129–130; effect of on
radical working-class reconciliation, 129; enthusiasm for, 129–
130
Fries Rebellion, Duane reports of,
153

Gallatin, Albert: on merchants, 115;
on small-producer republic, 113,
115
Galloway, Joseph, 31
General Loan Office, 85; credit terms
of, 85–86
George III birthday party, toast, 130

German Associators, democratic demands made by, 42
Girard, Stephen: bank owned by,
181; leadership during epidemic,
136, 137
Globe Glass Works, 100
Gookin, Charles, replaced as Pennsylvania governor, 22
Gray, John, xv; lectures by, 214;
compared to Blatchly, 215;
small-producer tradition views
of, 216; socialist writings absorbed by Heighton, 222; views
generally, 214–217
Graydon, Alexander, on revolutionary constitution, 53
Guardians of the Poor, 198
Guilds, as source of small-producer
tradition, 9

Hamilton, Andrew, attacks against by
Keithians, 24
Harbeson, Benjamin: as Constitutional Society leader, 54; as
craftsmen leader, 58
Harrison, Job: cordwainer career of,
176–177; dismissal demanded,
177; refusal to support strike,
177; reinstatement in Cordwainers' Society, 178; testimony
in cordwainers' trial, 176, 178
Hartz, Louis, on bank charter debates, 185
Hegemony, defined, 115
Heighton, William: call for working-
class movement, 222; campaign
of, xv; career, 222; critique of
caucus system by, 224; forms
mechanics' union, 228; historians' assessment of, xi; in politics, 230–232; lecture on
working-class unity, xi; on representative democracy, 223; publishes *Mechanics' Free Press*,
228–229; speeches by, 224–228;
universalism and, 225–226;
views of established clergy, 227
Heresy, cloth-producing districts as
centers for, 8

Hill, Christopher, on "layman's creed," 9
Hinshellwood, John, on postwar depression, 195
Hospitable Society: organized, 173; reports of, 174–175; work with laboring poor, 173–174
Howe, Gen. William, occupation and departure, 49
Humble, John, address of, 80
Hutchinson, Dr. James, 69; and revival of opposition to Federalists, 123–124, 125; as leading anti-Federalist, 82; as member of Democratic Society of Pennsylvania, 131; calls for town meeting, 127; death in epidemic of 1793, 136; defeated in 1780, 91; on Constitution and trade, 78–79; on third town meeting, 128; refusal to join Committee of Privates, 58; to Gallatin on 1792 election, 128; town meeting speaker, 55

Immigrants: as ingredient of new radical/working-class alliance, 124; during "youth" of laboring-class movement, 141; 18th-century, to Pennsylvania, 19
Imports, calls for restrictions on, 93–94
Independent Gazetteer: on Independence Day (1795), 138; on wealth and values, 103
Industrial revolution, in Philadelphia, 165–171
Inflation: continuing after Fort Wilson riot, 60; Revolutionary War, 47; wartime, 29
International trade, credit required for, 84
Ireton, Henry, response of to Levellers, 13
Irish immigrants, 124; effect of yellow fever on, 135; as part of opposition to Federalists, 124; craftsmen in, 124; radicalism among, 124–125; resumption of arrival of, 153
Israel, Israel: career, 145–146; Democratic-Republican candidate (1796), 143; elected sheriff of Philadelphia, 151; election of 1797, 147; founder of Democratic Society of Pennsylvania, 130; leadership during epidemic, 136, 137; loses seat, 148; popularity, 148; senatorial campaigns, 145

Jackson, Andrew, success of among workingmen, 231
Jay Treaty: as unifier of radicals and workingmen, 137–138; opposition to, 137, 138, 139
Johns, Mathew, seizure of, 59
Jones, Robert, hard money legislator, 21
Journeymen Cordwainers' Society, 176, 177
Journeymen's associations: benefits offered by, 118–119; daily functioning of, 118–119; English, 9–10; evolution of into trade unions, 120; organization of, 118
Journeymen's trade societies: absence of records of, 95; growth of, 221; militancy, 221; program, 96; rise of, 95; strikes by, 221
Joyce, Cornet George, 99
Junto: Benjamin Franklin's discussion group, 27; legacy of Leather Apron Club on, 27

Keith, Sir William: background of, 21–22; departure, 23, 27; election to assembly, 23; moderating influence of, 25; on debate with Quaker elite, 22; on wealth redistribution, 26; parable of the fat man, 26; resurgence of ideas of, 32, 33
Keithians, 23; alliance with city artisans, 22; rhetoric of, 24

Labor: as property, 5; theory of value in Benjamin Franklin pamphlet, 25

Laboring-class, divisions in, 171–172; *see also* Workingmen

Langhorne, Jeremiah, attacks against by Keithians, 24

Leather Apron Club (Tiff Club), 61; as predecessor to other organizations, 28; education from, 25; meetings by, 22–23; significance of, 26, 27; William Keith's connections to, 21

Leiper, Thomas, council candidate, 142

Levellers, 4, 10–11; demands encompassed in 1776 revolutionary constitution, 43; draft constitution of 1649 and, 11–12; echoes of in Keithian rhetoric, 23; economic program of, 12–13; influence, 14; leaders, 10; petition, 11

Levy, Judge Moses, as presiding judge in cordwainers' strike trial, 161

Libel, William Duane on, 154

Liberalism: compared to small-producer tradition, 103–114; contrasted to popular radicalism, 105; ideas of, 105

Leib, Dr. Michael, xv, 221, 230; death, 209; Democratic-Republican candidate (1796), 142–143; final attempt at coalition by, 199–200; introduces relief bill, 147; leadership during epidemic (1793), 136, 137; Old School leader, 192; Old School organization of, 181

Lilburne, John, 10

Lisle, John: Family Party of, 210; on Duane and Old School, 210

Lloyd, David: campaign against Penn's authority by, 18; death, 27; in creation of small-producers' coalition, 23

Logan, George: Democratic-Republican candidate (1796),

143; merchant-Republican leader, 183; on small-producers' republic, 113–114

Logan, James: as Penn's secretary, 18; attacks against by Keithians, 24; comparisons to Parliament made by, 24; on civil war themes in middle and lower ranks, 23–24; opposition to paper money, 20

London Corresponding Society, 151

London *General Advertiser*, 151

Machinofacture, 169–170; in textile industry, 169

Maclure, William: career, 204–205; educational experiments of, 204–205; on Gray's influence, 214

Madison, James: attacks on small-producer tradition, 105, 107; justification for inequality by, 107; view on liberalism, 105; views contrasted with conclusions of workingmen, 105–106

Manufacturing: industry types in Philadelphia, 167; Philadelphia as center for, 164–165, 166–167

Marsh, Joseph, on non-importation committee, 100

Marshall, Christopher: disowns radical movement, 58; Provincial Conference leader, 42

Mason, Sen. Stevens Thompson, leak of Jay Treaty by, 137

Matlack, Timothy: defeat (1782), 91; investigating Morris's finances, 57; leader of Constitutional Society, 58; Provincial Conference leader, 42; support for militia after Fort Wilson riot, 69; support of small-producer ideas, 64; town meeting speaker, 55

McClenachan, Blair, 125; as member of Democratic Society of Pennsylvania, 131; career, 126; Democratic Society leader after epidemic, 137; Democratic-Republican candidate (1796), 142; elected to Congress, 145; organizes anti-Jay Treaty demon-

stration, 126; organizes Democrats against Jay Treaty, 137
McCombs, Lawrence: establishes
Philadelphia Academy Methodist
Church, 172; Methodist minister, 172; rejects laboring-class
demands, 172
McKean, Thomas, 145; as merchant-
Republican leader, 183; elected
governor, 151; leadership of, 69–
70; letter from Matlack to, 69
Mechanics' and Manufacturers'
Banks, 183
Mechanics' Committee: demands for
economic reform from, 33; for-
mation, 70; organization of first,
33
Mechanics' Free Press, 214; contents
of, 229; started, 228–229
Mechanics' Union of Trade Associa-
tions, xi; formation of, 228
Merchant class: and non-importation
movement, 31–32; celebrated by
Pelatiah Webster and John
Witherspoon, 110–113; praised
by James Bayard and Harrison
Gray Otis, 115
Merchant-Republicans (Quids), 183;
repudiate Democratic-
Republican coalition, 183
Merchants: in West Indies trade, 75;
increasing wealth of, 39; profiteering, 55; profiteering in Revolution, 46; resistance to by
Associators, 47
Merchants and Manufacturers Society, 167
Merchants' Committee, 76; committee formed to petition Congress,
100; courting of workingmen by,
100–101; non-importation meeting called by, 98–100; on nationalization of trade, 76
Meschianza, on Howe's departure,
49–50
Methodists: organizational skills of,
217–218; origins of in Philadelphia, 217
Mifflin, Gov. Thomas: appoints

Dallas secretary of Commonwealth, 125; asks workingmen to
delay cannon salute, 139
Militia: broadsides after Fort Wilson
riot, 92; defeat in Fort Wilson
riot, 69; equalized participation
in, 47–48; in call for town meeting, 54–55; mobilization (1776),
45; seizure of Tory opponents by,
59; working-class support for, 28
Miller, Daniel H., defeated, 232
*Modest Enquiry into the Nature and
Necessity of a Paper Currency*
(pamphlet by Benjamin Franklin), 25
Monopolies, abolition of demanded
by Levellers, 12–13
Moral economy: in England, 62; in
Philadelphia, 63
Moral traditions, importance of English, on class formation, 3
*More Light Shining in Buck-
inghamshire* (pamphlet), 14
Morgan, Benjamin, Federalist opponent of Israel Israel, 147
Morrell, Capt., cavalry leader against
demonstration (1795), 138, 139
Morris, Robert, 127; campaign of
Paine for bank directed by, 91;
campaign to recharter bank directed by, 86; financial affairs in-
vestigated, 57; formation of Bank
of North America by, 84; orga-
nizer of Republican Society, 54

Nash, Gary, quoted on Penn's frame
of government, 16
National Bankruptcy Act (1800),
114
Nayler, James, Quaker founder, 15
Neff, Joseph: as radical academician,
205; closure of academy of, 205
Nelson, George, on popular discon-
tent, 61
Nevil, Thomas, Committee of Pri-
vates leader, 34
New Harmony, Indiana, 213
New Model Army, Associators' mili-
tia copied from, 28, 33

New School Democrats, 182; claim Jackson as candidate, 210; formation, 183–184; in bank debate, 184; interest in Owen by, 213; joined with Old in 1812 election, 192; pro-manufacturing views of, 190–191; resume conflict with Old, 199; success in election (1810), 188; success in election (1818), 200; views of, 189

New York City, as competing port, 165, 166

Nicholson, John, Roxborough manufacturing plan of, 166

Non-importation movement, Philadelphia merchants and, 31–32

Occupation, British, 48–51

Old School Democrats, 182–192; defeat (1810), 188; demise, 210; end of as a party, 200–201; focus on craftsmen, 190–191; in bank debate, 184; joined with New in 1812 election, 192; resume conflict with New, 199; views of, 189–190; views on banking, 185–186; Workingmen's Party platform compared to, 230

Otis, Harrison Gray, on side of merchants, 115

Overseers of the Poor, 118; as "Guardians of the Poor," 122

Overton, Richard: as Leveller, 10; thought of paralleled by William Keith, 22

Owen, Robert, 204; influence of ideas on workingmen, 213–214; socialist experiments of, 212; visit to Philadelphia, 213; writings published by Duane, 212

Paine, Thomas: and *Principles* (of Democratic Society), 132; as investigator of Morris's finances, 57; as pamphleteer favoring West Indies commerce, 76; as secretary of the Commonwealth, 58; campaigns for Morris's bank, 91;

on support for Bank of North America, 87; support for merchant hegemony, 113; support for small-producer ideas, 64

Panic of 1819, xv, 197

Paper money, as issue during 1722 depression, 20–21, 24

Party politics (*see also specific parties*), viewed by workingmen as futile, 103, 104

Patriotic Society, 51; composition of membership, 51

Peale, Charles Willson, 69; as investigator of Morris's finances, 57; defeated (1780), 91; refused to join reconstituted Committee of Privates, 58; retreats from politics, 70; support for small-producer ideas, 64

Penn, Thomas, 31

Penn, William: and "inner light" doctrine, 15–16, 17; and Pennsylvania colony plans, 16–17; colonial control by, 17; concessions granted by, 18; frame of government, 16

Pennsylvania Evening Herald, 125

Pennsylvania Packet (newspaper): on 1785 depression, 76; price control debate in, 63

Pennsylvania Society for Promoting Public Economy, 198

Pennsylvania Society for the Encouragement of Manufactures and the Useful Arts, formation of, 166

Philadelphia: population growth of, 38; trade fluctuations in, 38

Philadelphia Academy Methodist Church, established, 172

Philadelphia Typographical Society: constitution, 120–121; democratic character of, 122; organizational arrangement of, 122; organized, 120; ratification of constitution of, 121

Pinton, Dr. John, congressional candidate, 187

Politicization, tradition of popular, xii

Poor relief, cost (1780s), 116
Popular radicalism: history of leaders and followers of, 69–70; rebirth of, 103–109
Powel, Samuel: as alderman and mayor, 89; establishment leader, 88; Federalist chairman (1792), 127
Price regulation, 51, 193; by cabinet and chair-makers, 119; of coffins and burials (1793), 136
Price-regulation movement, 57 (see also Price-control movement); committee of 13 and, 65; conflict over, 63; failure of, 64; history of, 62; importance of, 57; opposition to, 63–64
Price-control movement (1779), xiii
Princeton, Battle of, Associators in, 46
Principles (of Democratic Society), 133; appeal to workingmen in, 133
Printers: association of, 119; evolution of into trade union, 120; strike by, 119, 120
Producers, defeat of, 89–90; see also Small-producer tradition
Property: as issue in revolutionary constitution, 44; challenged in Declaration of Rights, 44; concept of, 65
Property rights: Madison on, 106–107; workingmen's view of, 106
Proud, Robert, quoted, 45
Provincial Conference: call for constitutional convention by, 42; leaders of, 42
Prynne, William, Levellers' rise to oppose, 10
Public education, free, support for, 230
Pullis, George, as leader of Journeymen Cordwainers' Society, 178–179
Putney debates (1647), as ending Levellers as a party, 13

Quaker Yearly Meeting, seizure of Tories at, 59
Quakers: and resistance to military obligation, 45, 46; as minority faith by early 18th century, 19; decline of egalitarian nature of, 19; development of beliefs of, 15; harassment of wealthy, 93; purges of, 15; relationship of to Ranters, 15; Test Act effect on, 74
Quids, 183 (*see also* Merchant-Republicans, New School Democrats); organize bank, 184; renamed New School Democrats, 183–184

Radical-popular movement, 45; breakup of, 91
Radicals, opposition movement and, 126–128
Ranters, beliefs of, 15
Rawle, Francis, in formation of small-producers' coalition, 23
Read, James, accused of election intimidation, 149
Reed, Joseph, on Fort Wilson riot, 60
Religion: democratic aspirations expressed in, 14; universalist dogma and small-producer tradition, 131
Report of the Library Committee, 198
Republican Society, 54; call for new state constitution by, 72
Republicans: attacks by on constitution, 73; attempts in legislature to change constitution, 74; attention toward workingmen's issues, 98; attitudes of laborers toward, 96; campaigning for laboring-class support, 94; domination by, 91; exclusion of artisans from leadership, 92; for constitutional convention, 77; goals, 71; in election of 1782, 72–73; in post-Revolutionary War years, 70; opposition to Constitutionalists, 70–71; program, 72; protection-

Republicans (*Cont.*)
ism, 76; support for new state
constitution, 72
Revolutionary movement, craftsmen
in, 39, 40
Reynolds, Dr. James: and beginnings
of socialism, 202; prosecuted under Sedition Act, 202; writings,
202–203
Rittenhouse, David, 69; as founder of
Democratic Society of Pennsylvania, 130; as investigator of
Morris's financial affairs, 57
Roberdeau, Gen. Daniel, town meeting speaker, 55
Robinson, James, on Federalist election threats, 149
Rodney, Caesar A., defense counsel
for cordwainers, 162
Rural-urban friction, 43
Rush, Benjamin: as Republican
spokesman, 70; coverted to Republicanism, 91; disowns radical
movement, 58; first diagnosis of
yellow fever by, 134; on revolutionary constitution, 53–54;
Provincial Conference leader,
42
Rush, Richard, guides Owen, 213

Sadler, Matthias, as council candidate, 142
St. Clair, Arthur, endorsement of,
127
St. George's Methodist Eposcopal
Church: divided by class factions, 172–173; schism, 174–176
Schlosser, George: as Constitutional
Society leader, 54; as craftsman
leader, 58
Second Anglo-American War (War of
1812): declared, 191; prosperity
resulting from, 192; soldier suffering in, 193
Sergeant, Jonathan Dickinson: as
member of Democratic Society
of Pennsylvania, 131; careeer,
125–126
Second Great Awakening, 217

Servants, declining ranks of indentured, 40
Seven Years' War: aftermath, 31;
economic changes caused by,
29–30; experience of Philadelphia laboring class in, 29
Shippen, Peggy, wife of Benedict Arnold, 155
Shipyards, unemployment in, 196
Simpson, Stephen: as Old School
leader, 185; career, 208; leader
of Old School remnants, 201; on
banking issue, 185–186; on 1810
election coalition, 181–182; on
split, 187; protest leader, 210;
socialist writings of, 207–209;
views on manufacturing, 190–
191
Sims, Buckridge, seized, 59
Small-producer tradition, 4–7; anti-
Federalists favor, 115; appeals by
Democratic-Republicans to
(1796), 142; breakdown of, 171;
Democratic-Republican Party
speaking to, 128–129; Duane
writings on, 154–157; English
background of, 7–10; extended,
141; first public expression of,
10; Franklin in, 25; Gallatin vision for, 113, 115; ideas in, 64;
in Gray's theories, 216; in Mechanics' Union of Trade Associations, 228; in socialism, 209,
in Universalism, 131; in
Workingmen's Party, 231; intellectual crisis for, 107; Levellers
and, 10–13; Madison attacks on,
105–106, 107; Old School attempts to return to, 201; popular
participation tradition in, 17; re-
assessed in light of radical ideology, 150; role of religion in, 14–
15
Smilie, John, on banking and credit,
86
Smith, Adam, influence of, 63
Social equality, tradition of, 6
Socialism: as middle-class idea, 211;
beginning of, 201–202; Blatchly

and, 205–207; critiques of capitalist society in, 220; Duane contribution to, 203–204; Heighton on, 221–222; ideology of artisan, 211; Maclure and, 204–205; Reynolds writings on, 202–203; Simpson writings on, 207–209
Society for Constitutional Information, 133
Society for Political Enquiries, 133
Society of Cincinnati, 139
Sons of St. Crispin, 162
Speakman, John, plan for Owenite community by, 213
Stamp Act, 31
State constitution, 37
Story, Thomas, seized, 59
Strike, first recorded in Philadelphia, 119, 120; *see also* Cordwainers' strike
Suffrage: exclusion of craftsmen from local by restored charter, 89; in 1776 constitution, 42; universal male, 37
Supreme Executive Council, trial of collaborators by, 51
Swanwick, John: elected to Congress, 145; elected to state assembly, 128

Tariff bill: meetings proposing, 100; passage (1785), 99
Tariffs, party and class attitudes toward, 96
Test Act, 74
Textile mills, 165
Textiles, 169–170
Thackera, James, 230
The Committee (in epidemic of 1793), 136
Thompson, Edward, on markets, 62
Thompson, William, 214
Thomson, Charles, campaign in favor of non-importation, 3
Thorne, William: Committee of Privates leader, 34; Constitutional Society leader, 54

Tories, extravagance of, 49–50
Town meetings (1792), 126, 127
Townshend duties, community protest against, 3
Trade: deterioration of international, 165; postwar, 195
Trade unions, evolution of journeymen's associations into, 120
Triumvirate of Pennsylvania (pamphlet), 24–25
Troost, Dr. Gerard, plans for Owenite community by, 212–213
Typographers (*see also* Printers, Philadelphia Typographic Society), organization of, 123

Umemployment: in 1820, 168–171; of women (1820), 169–170; post-Second Anglo-American War, 195–196, 197, 198; post-Seven Years' War, 38
United Irish independence movement, 124, 125; Bache support for, 152
U.S. Constitution: commercial clauses discussed, 78; ratification in Pennsylvania, 82, 83; Republican views of, 77–78
Universalist Church: doctrines, 131, 225–226; support for Heighton from, 225

Voting patterns, impossibility of tracing, 91

Wage labor force, formation of, 40
Walwyn, William, 10; on poverty, 10–11
War (*see also specific wars*), effect on Philadelphia working class of, 29
Washington, George: arrives in Philadelphia, 109; celebrations honoring, 109
Watson, John, on Methodist schism, 173
Wealth, distribution of, 39
Wealth of Nations, influence of, 63
Webb, Thomas, 217

Webbe, John: declaration of principles by, 28; "Z" letters, 27–28

Webster, Pelatiah: against fixed prices, 112; favors unregulated commerce, 112; fiscal views of, 111; merchant as hub in writings of, 110–113; on happiness, 112; *Political Essays* by, 110

Wharton, John: as establishment leader, 88; on council, 89

Whitefield, George, revivals compared, 217, 218

Williams, William, on wartime prices, 47

Willing, Thomas, organizes Republican Society, 54

Wilson, James: and Fort Wilson riot, 59; chairs town meeting (1792), 126–127; organizes Republican Society, 54; wealth of, 60–61

Winchester, Elhanan, 131

Winstanley, Gerrard: on the poor, 32; thought paralleled by William Keith, 22

Winthrop, John, Jr., as "Joyce, Jr.," 99

Witherspoon, John: *Lectures on Moral Philosophy* on merchant class, 110; opposes setting prices, 112

Women: economic plight after Panic of 1819, 198; in textile production, 169–170

Working-class movement: as part of politics, 4; creation of, xii; party affiliations in, 92; political presence after 1779, 92; propaganda on, 93

Workingmen: association advocated for, 101; economic condition in 1780s, 116; finding common cause with others in Revolution, 44–45; in elections, 95–97, 98; independence of, 101; independent political stance urged for, 97; living conditions of, 172–175; organization of, 94; outlook toward merchant class by, 95; return to politics by, 149; vote-splitting in 1810, 187

Workingmen's Party, xi–xii; call for by William Heighton, xi; demise of, 232–233; founded (1828), 229; history of, 231; in 1829 election, 231–232; platform, 230

Yellow fever epidemic: death of Dr. Hutchinson, 136; impact of 1793 on laboring class, 134–135; in 1797, 146

Young, Thomas: as Provincial Conference leader, 42; moves from Philadelphia, 58